Lost in the Battle for Hong Kong

December 1941

The front cover of this book is dedicated to the memory of Ruth Baker (née Sewell). Sadly, Ruth passed away on New Years Day 2019, but her memory remains with me in my memoir. She provided the photo of the Bowen Road group of children used on the cover and at the end of chapter 7.

Lost in the Battle for Hong Kong

December 1941

Bob Tatz

A MEMOIR OF SURVIVAL, IDENTITY AND SUCCESS. *1931 – 1959*

Lost in the Battle for Hong Kong! December 1941

A Memoir of Survival, Identity and Success 1931 – 1959

My thanks to all those who kindly permitted reproduction of photos, images and text, to enhance my narratives. This has proved invaluable to assist my memory and recall events, especially in the early years. Any omissions to acknowledge the above is unintentional, and I would welcome notification through PageMaster to redress any concerns.

Published by Bob Tatz, Edmonton, Canada

ISBN: 978-1-77354-125-9

Publication assistance and digital printing in Canada by

Eilish

FOREWORD

The Real Empire of the Sun: The Remarkable Story of Bob Tatz

On December 8th, 1941 the Kowloon waterside was packed with men and women fleeing the Japanese army which had begun its assault on Hong Kong early that morning. The colony was home to over a quarter of a million refugees from the war in China and no-one was in any doubt as to how violent the Japanese could be. Few people in the middle of all that fear and panic would have noticed a boy not yet in his teens, completely without adult help and protection, making his own way to the temporary safety of Hong Kong Island.

Hong Kong was at the start of eighteen days of fighting, to be followed by three years and eight months of brutal occupation. The historian Tony Banham has estimated that this was to cost over 300,000 lives, roughly one in six of those living there at the time of the attack. Who would have guessed that the ten year old boarding the ferry on his own would be one of the survivors?

Even before that dreadful December day, Bob Tatz had experienced much more than the normal ration of loss. Death had taken his father, stepfather and his mother, and he no longer saw the 'beloved' Chinese amah who had once helped care for him. He did have two older sisters and a godmother somewhere on the island, but he had no reason to expect they would come looking for him. He was on his own.

When J, G. Ballard came to write the classic novel of civilian internment *Empire of the Sun*, he decided to make his child self – 'Jim' – experience Shanghai's Lunghua Camp as an orphan, even though in reality Ballard was there with both his parents. Stephen Spielberg's film underlined the drama and poignancy that this decision allowed Ballard to create, and we find plenty of those two qualities in Bob Tatz's memoir - but in this case we know we are reading sober historical chronicle not imaginative fiction.

Bob notes in his introduction the fascination that his story has long held for historians of the Hong Kong war, and this group will not be disappointed by his account. This memoir throws light on a host of issues of importance to those who study the period. But in no way is it just for historians. It tells a story that will intrigue everybody interested in the response of the human spirit to circumstances of the utmost difficulty.

Bob's physical survival was, of course, most at risk during the eighteen days of bitter fighting. He was lucky that soon after boarding that ferry on December 8th he was taken into the care of William Sewell, a Quaker missionary and university chemist. His temporary adoption by the Sewell family meant that he was surrounded by an interesting group of adults and children, and what follows makes an exciting tale of life under bombardment. We also learn of his group's reaction to the eventual arrival of the victorious invaders. Some readers will be surprised at the reasonableness of most of the Japanese who interacted with Bob and his party. The nightmare that was beginning was not due to the fact that the occupiers were uniformly lacking in humanity: the problem was that the inhumane ones were in command. As the war continued, the Kempeitai – almost universally referred to as 'the Japanese Gestapo' – tightened their grip on Hong Kong, and Bob's wartime life took place in the context of the terror they inflicted on the population. Luckily he found two places of safety.

The first was Stanley Camp, set up by the Japanese authorities and run largely by its British inmates. Like most Allied civilians Bob was sent there early in 1942, and his account of camp life has surprises of its own. His picture of the relatively good situation of children there is borne out by other sources, although the freedom to skip classes without adult interference seems to have given him a uniquely pleasurable time. Inside the barbed wire everyone was protected from the worst brutalities of Japanese rule, but it still takes one aback to read that during his roughly five months in Stanley a ten year old boy without an adult protector, could have begun to find 'serenity'. At such a moment we come to understand that, although the author's lonely childhood had left him somewhat diffident and lacking in self-confidence, it had also given him the capacity to make the best of things and to create his own satisfying world from whatever elements lay to hand.

In the early summer of 1942 Bob was sent from Stanley to his second place of refuge, the Canossian Convent, where he joined a variegated and colourful group being cared for by the Italian nuns. This decision was never explained to him and left him furious at what he perceived was his rejection by the British authorities. It would probably have been small consolation that the move would one day enable him to write what is to the best of my knowledge the only detailed English-language account of life in the Convent seen from a 'refugee' perspective! In addition, Bob's location gave him splendid views of the American air raids on the harbour and he provides us with some valuable descriptions of these. When the Allied fleet finally arrived in Hong Kong - more than a fortnight after the Japanese surrender – Bob was perfectly placed to go down to the Naval Dockyard and begin to make up for the years of 'emotional deprivation' by taking up the attention of the sailors - and in due course to give us a unique child's eye view of the reoccupation.

I've begun by discussing Bob's account of his wartime experiences because his tale is rare and fascinating and undoubtedly makes up the highlight of the book. Nevertheless it would be a mistake to focus on this section alone: the narratives of what came before and what was to follow are valuable in themselves and help us understand the full import of Bob's story.

Hong Kong before the war was status conscious to a degree that it is now happily hard to imagine. Bob's family belonged to a category he calls 'undefined European expatriates' – nationals of states without much institutional presence in Hong Kong. This was a group on the edge of 'European' society and both historians and interested general readers will be grateful for a rare glimpse inside a world about which too little is known.

Bob's account of his early life, marked by multiple tragedies as it was, makes it clear that he was nourished by both his mother's love and the care provided by his amah, Ah Kai. Amahs played an important role in the household of many expatriates, but Bob's experiences in the market of the now demolished Walled City in Kowloon and with Ah Kai's friends and family in the countryside seem to have been unusually extensive and to have fostered a lively interest in his surroundings, while helping to make up for what was absent in his family life. It is interesting that even as a young child he grasped something of the racial hierarchy of pre-war Hong Kong, a society in which the Chinese majority were subject to systematic discrimination. Things got even worse under the Japanese and it seems likely that Ah Kai was amongst the huge number of Chinese who did not survive the occupation. Many British were moved by the tragedy that played out before them, and some were amazed and rather ashamed when Chinese friends or servants somehow scraped together the money to send food parcels into

Stanley, but sadly the good intentions that the war fostered did not end the racist system, although it did soften it somewhat.

In any case, I think that the love of his mother and Ah Kai helped create a resilience that allowed him to survive the horrors of war and occupation and a psychological strength that enabled him to overcome the problems of what one writer has called 'the war after'. For some people the liberation that arrived with the British fleet in August 1945 did not bring a return to normal life: wars, to quote Tony Banham again, have long tails, and in some tragic cases peace brought madness and suicide. The final section of this memoir, as well as providing rich material for those interested in the maritime history of the Far East, shows us how the author found the perfect setting in which he could get back on track psychologically: the merchant navy provided him with new and colourful experiences in the context of the discipline, hierarchy and fellowship that enabled him to complete his emotional education in the same way as studying engineering brought his interrupted schooling to a successful conclusion. I'm sure many readers will be moved by the way in which that scared and abandoned orphan boy eventually found fulfilment in work, love and parenthood.

The memoir ends with the author deciding to sacrifice a secure financial future with Jardine's and leave Hong Kong for Canada, a country in which he had no family, friends or other contacts. This willingness to gamble everything he had achieved on a leap into the unknown is powerful testimony to the fact that, whatever happened in his early life and in the three years and eight months of the Japanese occupation, Bob Tatz grew into a man with the courage to trust his own judgement. This memoir – written with a remarkable absence of bitterness – charts his path through a tumultuous childhood to self-confidence and maturity.

Brian Edgar
Formerly Professor of English Language and
Literature at the University of Yunnan'

Contents

List of place names mentioned that have changed since the events of this story

Madras – Chennai (India)
Calcutta – Kolkata
Bombay – Mumbai
Chittagong –in East Pakistan, now Bangladesh since 1971

Pusan – Busan (South Korea)
Port Swettenham – Port Kelang (Malaya)
Jesselton – Kota Kinabalu (Malaya – Borneo)

Burma – Myanmar
Bassein – Pathein
Moulmein – Mawlamyine
Rangoon – Yangon
Verkhneudinsk – Ulan-Ude
Saigon – Ho Chi Min City
Indo-China – Vietnam

Ceylon – Sri Lanka
Peking – Beijing
Tientsin – Tianjin
Canton – Guangzhou

Victoria City (Hong Kong) – Central (District)

"We do not receive wisdom. We must discover it for ourselves after a journey through the wilderness which no one else can make for us, which no one else can spare us. For our wisdom is the point of view from which we come, at last, to regard the world."

Marcel Proust – *Remembrances of Things Past*

"There is pleasure in the pathless woods,
There is rapture in the lonely shore,
There is society where none intrudes,
By the deep sea, and music in its roar:
I love not man the less, but Nature more,
From these our interviews, from which I steal
From all I may be, or have been before,
To mingle with the Universe, and feel
What I can ne'er express, yet cannot all conceal"

Lord Byron – *Childe Harold's Pilgrimage*

INTRODUCTION

Early historical records of 1942 indicate the presence of a 10 year old boy named Robert Tatz, listed as a civilian internee in Stanley Internment Camp.

"Who is this orphan boy?" everybody asked.

About 70 years later, this "orphan boy", by now an octogenarian and a grandfather, was questioned in Hong Kong by several historians to help solve this mystery. Even to this day, many important questions about his appearance at the time remain unanswered.

These memoirs are the result of delving into this mystery. I reached as far back as memory allowed to before mother's death, who was the remaining surviving parent. I then endured loneliness for two and a half years in a boarding school right up to the beginning of the Battle for Hong Kong. I had turned 10 years old only two months earlier. Paradoxically, I was not at all unhappy at the break in my lonely life, as now I was associated intimately with a group of people collectively seeking shelter. This led me to the gates of Stanley Internment Camp.

Still largely on my own, I was approaching the age of adolescence, which led to a period of self-discovery, by the end of which time I was well on the road to entering the world on my own merits. An engineering apprenticeship, engineering education, experience with the Hong Kong Auxiliary Air Force, all contributed to credibility in Hong Kong expatriate society. The greatest satisfaction of all was my subsequent career with Jardine's shipping fleet, the Indo-China Steam Navigation Co. Ltd., which included a stint with the Ministry of War Transport (MOWT) in Korea.

At the end of it all, survival turned into sweet success, despite the unique challenges encountered in the earlier period of my life. This is my story.

This story is set in Hong Kong, but it is not a history about Hong Kong. It is my personal story of survival from the time I was orphaned at seven years old, and the challenges I had to face on my own. As all children do, I relied on timely support of adults until I was old enough to fend for myself. Many of these adults failed me, but others were kind. I dedicate this book to everyone who ensured my survival, took an interest in my welfare, was part of a happy memory or helped to give me the experiences that made me who I am today.

CHAPTER 1: BEFORE THE BEGINNING

Exodus From Troubled China

1920s

The China my European parents arrived in separately in the early decades of the twentieth century was a nation shaped by centuries of internal political change, European exploitation and trade, and unparalleled population and economic growth.

The Canton System of trade, implemented in the eighteenth century, had restricted foreign traders to specific ports so that China was in control of her own imports and exports. Europe's appetite for Chinese goods was insatiable, and foreign traders, especially the British, resented the restrictions imposed on them. Only by making China as dependent on opium as Europe was on Chinese goods—and fighting two wars over the opium trade—did foreign traders loosen China's hold on its ports. Defeated in military actions, China was obliged to sign a series of treaties with foreign powers. These agreements, known in China as the Unequal Treaties, forced China to pay large reparations, open new trading ports, cede or lease territory to foreign powers, and grant concessions such as extraterritoriality to Europeans living in China.

Chinese resentment of this treatment at foreign hands gave rise to the anti-foreign, anti-Christian Boxer Rebellion between 1899 and 1901. Quashed again by an international force led by the British, China signed a humiliating peace agreement, and the Emperor and Empress fled Peking for the provinces. This set the stage for the end of the imperial dynasties and began the era of the warlord government, the Kuomintang, and the rise of the Communist movement under Mao Zedong.

My father, Kalman Tatz, was born in Hungary on April 30, 1893, and held Austro-Hungarian citizenship under the Habsburg regime. He grew up in a period of rapid change, high tension and diplomatic clashes. He was 21 when the First World War started in 1914. Records of his life are scarce, so it's not clear how the politics of the day or the outbreak of war affected him. He might have been worried about mandatory conscription or pressured to take immediate action. With Europe on the brink of chaos, Russia in no better situation and China in disarray, he had few options. Many questions about his life as a young adult remain unanswered. I know that he was in Peking in 1917; he was an artist, and a signed and dated watercolour painting places him there at that time.

My mother, Antonina Shangin, was born in Verkhneudinsk (now called Ulan Ude) in Siberia, on February 17, 1904. Her father was a career medical officer in the Tsar's army who, because of his position and loyalties, was also a landowner. Antonina's brother also served under the Tsar at the time of the 1917 revolution and the subsequent savage civil war. The Bolsheviks executed both father and son, and confiscated the family's property, prompting Antonina, her mother and two sisters—all that remained of the family—to flee east into China. They eventually settled in Harbin around 1920.

Harbin was the headquarters of the Chinese Eastern Railway (CER), which the Chinese and Imperial Russian governments built across Manchuria to connect the Trans-Siberian line with Vladivostok. Harbin grew into a sophisticated European-style city between 1903 and 1917, and the city's Russian population tripled by 1922 due to the cataclysmic events in Russia during that time.

With the Bolsheviks firmly in control over all of Russia, China signed an agreement with the Soviet Union in 1924 for joint management of the CER. By this agreement, only Soviet and Chinese citizens were permitted to work on the railway. This meant the Harbin Russians had to choose not only their nationality, but also their political identity. From necessity, many Harbin Russians took Soviet citizenship, but there were also Harbin Russians who remained stateless, then became unemployed.

The Shangin family was caught up in this second wave of Russians seeking refuge further south in China or immigration to Australia, the United States or other countries friendly toward displaced persons.

§

My father was an instructor at an art college in Harbin; my mother one of his adult students. He was 30 and she was 19 when they married on October 31, 1923. By marriage, Mother was no longer considered a Harbin Russian or displaced person. My sister Margaret was born in Harbin on March 6, 1926.

Father had to have been a qualified artist to secure a teaching appointment at an arts college in Harbin. He also held exhibitions of his work in Peking, though the 1917 watercolour is all that survives of what must once have been a sizeable portfolio. There is also some evidence that he produced bronze sculptures of certain warlords in northern China.

My parents were familiar with and lived in areas governed under treaty in Tientsin (Tianjin), Harbin, Peking (Beijing) and Canton. Surviving photographs show evidence of a comfortable family life in northern China: fine furniture, servants and an easy, affluent lifestyle. Photos from their life in Tientsin and Shameen corroborate a notably well-off life, with some social connections in diplomatic and expatriate circles.

The turmoil of the Chinese civil war and rising Japanese aggression probably added urgency to the family's plan to move out of northern China. I have no exact dates, but in 1928, the family boarded a ship in Ningbo that was probably bound for Canton.

In Canton, they lived in the fashionable foreign concession of Shameen, where my sister Julie was born in April 1930. From that base, Father might have made a few excursions into Hong Kong before the final move. Foreigners and Chinese were moving to Hong Kong in great numbers. At the time, it was considered the safest place in the region—a British colony immune to China's internal conflicts and regional tensions.

§

British colonial policies and attitudes kept Hong Kong racially segregated and politically polarized. Race laws like the Peak Reservation Ordinance prevented the Chinese—even British-educated upper-class Chinese—from living in areas like Victoria Peak, which were reserved for the white elite.

By the time my parents arrived in 1931, Hong Kong had a decidedly British flavour: its parade grounds, barracks, clubs, Royal Naval Dockyard, European-style hotels, government offices, the governor's residence, banks, a university, a major hospital and other office buildings and shops. Victoria Peak dominated the western portion and was connected to Central by a funicular cable car system. The eastern portion was filled with Chinese shops; there were crowded markets in Wanchai, the Causeway Bay district and its typhoon shelter, the Happy Valley Race Course adjacent to the Colonial Cemetery and many hole-in-the-wall Chinese restaurants. Development in Kowloon included fewer commercial buildings and more residential units, many with small garden plots.

Of the 850,000 residents at the time about five percent were British, European, Australian, Canadian and American expatriates who managed merchant houses, financial institutions, shipping, engineering, hospitals and health care, education, and diplomatic representations. Armed forces were divided into officers and other ranks, but also with similar conditions and restrictions as for civilians. There was a growing class of missionaries, either located in Hong Kong or in transit to various destinations in the interior of China.

The largest population group in this melting pot of cultures and traditions was the Cantonese. A small number of upper-class Chinese were accepted in the Hong Kong hierarchy, mostly to serve as liaisons and mediators between the colony's government and the Chinese population. The government, merchants, and foreign traders depended on these Chinese middlemen, known as compradors, in the development and management of commerce. By and large, however, the Chinese supplied the labour, and were treated as at the bottom of the social order by the whites. Politically, the majority Chinese population also had little to no official governmental influence throughout much of this early period.

Hong Kong society was loosely structured along these lines when I was living there:

- Expatriates (include the ruling and merchant classes).
- Chinese wealthy elite (the "compradors").
- Eurasians with strong local connections.
- Native Chinese (the largest demographic group).
- European refugees (many transients).
- Undefined European expatriates (the smallest group who often had no consular representation, or institutional robustness [churches, schools, stores], but with time they could move into position within the ruling and merchant classes). My family was part of this group.

Because of strong home-country representation, the ruling and merchant classes were the most favoured segment of society. For the opposite reason—weak social and political representation by country of origin—undefined European expatriates were the least favoured.

In the British colony, whites enjoyed the advantages of social status, but wealth and power were more tied to commerce.

Most expatriate households—even those of undefined European origin—employed one or more domestic servants. These servants lived in segregated quarters at the residence of the expatriates and were on call 24 hours a day. In addition to housekeeping, they also cared for preschool-aged children while the parents pursued an active social life.

§

When my parents and sisters arrived in Hong Kong from Canton, they adjusted without difficulty to a colonial lifestyle not unlike the one they left behind. The major difference was that Hong Kong was a British Crown colony administered by the Colonial Office in London under British rule of law and not a concession area or a treaty port. As part of the British Empire, however distant from the centre of the empire, Hong Kong felt more protected and secure than anywhere else in China.

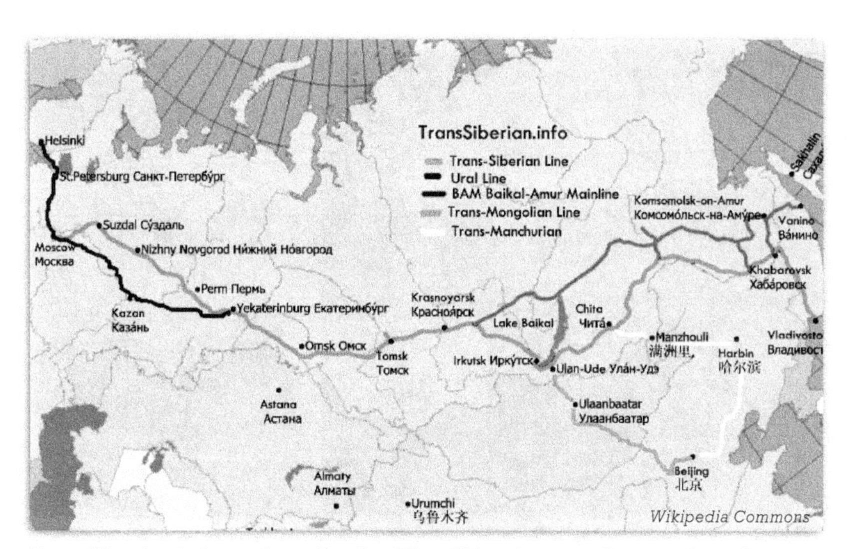

TransSiberian information, showing Ulan-Ude where mother was born in Siberia, and Harbin where her family took refuge from Russian Civil War.

Family picnic in Shameen. LR: Margaret, Ah Kai, Papa, a friend, and Mother. c1930

Father in Tientsin (Tianjin) or Shameen (Canton). c1929

Mother in Harbin. c1920.

CHAPTER 2:
LIFE IN HO MAN TIN

My Early Years

1931 – 1937

The family arrived in March 1931 and settled at 98 Argyle Street in Ho Man Tin, near the intersection of Argyle Street and Prince Edward Road, not far from the Old Kowloon Walled City. I was born in Ho Man Tin (Hong Kong) on October 11, 1931.

There is no evidence to suggest that Hong Kong was my parents' destination of choice. It turned out to be a welcome respite, and life was going well when the fates conspired to ground the family.

It was around this period, when I was still a toddler, that Olga Robinson first made an appearance in my life. She told me many years later that, during one of her visits to the flat, I had plied her with 'gifts' of all movable objects within my reach, and happily accumulated them at her feet. At the time, she was highly amused and overwhelmed by my 'generosity'. Many years later, she became my godmother and guardian.

In November 1932, one year and eight months after arriving in the colony, Father died. The official cause of death was 'pulmonary tuberculosis cardiac failure'. Tuberculosis is an infectious disease of the lungs. It was rampant in China in the 1930s and is still significant. The official cause of death implies that the cardiac (heart) failure was related to or a result of tuberculosis, but other factors might also have played a part. He had spent hours in a hot, smoke-filled foundry, smelting and casting bronze sculptures. Like millions of others, he was a heavy cigarette smoker. These additional assaults must have taken their toll on his heart and lungs. He was 39 years old when he died.

§

Mother was left a single parent, at the age of 28, to raise three young children on her own. Margaret was six, Julie two-and-a-half, and I a month past my first birthday. The prospect would have been daunting, and she must have relied heavily on friends and our long-time servant, Ah Kai, for help. It was then that my oldest sister, Margaret, was placed in the Canossian Convent boarding school on Caine Road, and Julie eventually followed her when she was old enough to be accepted. I used to see my sisters only when they came home on weekends or during school holidays. These early and lasting separations from my sisters would become the norm in our lives. I have no real memories of Margaret in those early years, and only a few of Julie.

This freed Mother to initially operate a gift shop located in the Peninsula Hotel, where she sold her landscape sketches, jewellery, postcards of her own design, silk flower arrangements, and sundry novelty items. With Father gone, supporting a family on her own meant spending a lot of time at the shop. I remember visiting the Peninsula Hotel with my mother, but I don't think she kept her shop there very long. Eventually, she moved this operation to a space at Estelle Dress shop on Mody Road in Kowloon, not far from the Peninsula Hotel. The dress shop was owned by a friend named Mrs. Karpusheff. I remember Mrs. Karpusheff well as a friendly and kindly person.

§

Mother sometimes took us to the beach at Kowloon Bay in Hung Hom. Not far into the bay was an anchored floating pontoon deck where swimmers would jump off into deeper water. The raw sewage pumped into the bay made it unsafe for swimming, but the beaches and the view were beautiful, and we played in the sand and splashed around in the water. The exposure to untold toxins may have boosted our immune systems, because all my siblings are still alive to this day, and we suffered no ill effects from the pollution except the time Julie nearly lost her leg from an infection.

Earlier in that fateful day, Julie had nicked the skin of her knee while cutting pictures out of magazines with a pair of sharp nail scissors. Neither she nor anyone else noticed the tiny puncture wound. We went to the beach as planned, and she picked up an infection from the polluted water. She had to be hospitalized and nearly lost the limb due to the severity of the infection.

Kowloon Hospital was a short distance from home, so Mother walked, and I peddled my tricycle beside her. I remember examining Julie's left knee and seeing the scars from the operation when Mother and I visited her in hospital. On the way home, I nearly joined her as a patient when I took my feet off the pedals and attempted to free-wheel

An outing with Ah Ngan and my sister Julie. This was in Ho Man Tin. c1937

Me at about 4 years old waiting for Mother at Day Care. c1934.

At Hung Hom beach with my sisters, Julie and Margaret. A family outing with mother who snapped the picture. In the background is a mat shed used as change facilities, and further back is the Hok Un Power Station. c1937

my tricycle down the slight hill. With no resistance from my feet, the pedals spun so fast I couldn't slow down. I lost control, was flung off the trike and tumbled along the recently resurfaced road. The fresh layer of sharp-edged gravel chips atop the asphalt scraped every bit of exposed skin. Badly frightened and bleeding, I bawled my head off. My mother scooped me up and took me back into the hospital, where they cleaned me up, good as new, and sent me on my way with a lollipop. I walked home, as Mother pulled the tricycle behind.

§

When Julie and I got along, we often amused ourselves by playing with a windup gramophone in the house. A favourite record was music of "The Blue Danube" waltz by Johann Strauss. To control the volume, we would stuff an old sock or a rag into the amplifier cavity to reduce the sound, and when the waltz slowed to a dirge, we hurriedly cranked it back up to speed. But we didn't always get along, and because our mother was so often away from home, we were left to resolve any disputes ourselves. Julie was older, bigger and accustomed to having her own way, and conflicts between us seldom resolved in my favour.

I vividly remember one occasion (I do not recall the incident that sparked it) when she was determined to get her own way with me.

"You'll do as I say!" Julie insisted.

"I won't!" I retorted.

She seized my most valuable toy and held it up. "I'll break it then," she declared grimly, "unless you listen to me."

"I won't!" I repeated.

She began to twist and distort it.

"Stop!" I pleaded."

"Then do what I say!"

By the time I finally acquiesced to her demand, my toy car was wrecked beyond usefulness or repair. That left me in tears.

When, as adults, I told her about this incident, Julie just laughed. She doesn't remember the event and the reminder meant little to her; she only relished the idea of having triumphed over me. The experience was formative for me, and it taught me that "possession is nine-tenths of the law".

Mother and Uncle Nicolas

Nicolas was a former cavalry officer of the Tsarist regime, and a friend who knew my parents when they were living in Canton. When Father died, Nicolas was already an established friend of the family. He was the proprietor of the Victoria Riding School located in Ma Tau Wai, not far from where we were living. The school offered horse riding lessons to the public and also rented horses to expatriates who hunted wild boar under the shadows of the nearby Kowloon hills.

Two entries in the *South China Morning Post* (SCMP) newspaper are of special interest:

> **June 7, 1934** *Page 10, The forthcoming wedding is announced of Mr. Nicolas Alexandrovitch Rojdestvin, of the Victoria Riding School and Miss Antonina Tatz.*

> **June 30, 1934** *Page 9, The wedding took place at 11 o'clock on Thursday morning at the Registrar's Office of Mr. Nicolas Alexandrovitch Rojdestvin, riding master at the Victoria Riding School, and Miss Antonina Tatz. The wedding was witnessed by D. Glazoonoff and G. Tarastjerna.*

About two years after Father's death, Uncle Nicolas became my stepfather; Mother was 30 and Nicolas was 40. I was about three years old at the time. I have a very vague recollection of this event, and the only connection I could make was Mother coming to fetch me home from a child care facility, after they presumably had returned from their honeymoon.

To me, he was the only father figure in my childhood, and I adored him. I think I was taught to call him Uncle Nicolas, which is a normal courtesy from children for any male friend of a household in Hong Kong. From then onwards, I found myself increasingly in the care of Ah Kai, while Nicolas was busy with his riding school, and Mother was busy with her small gift shop. In between, they were busy with each other and enjoying life. It was two years later, when I was about five, that I began visiting the riding school; this was probably in late summer or autumn of 1936.

Uncle Nicolas was gregarious and social. My mother sometimes took me to visit him at his work in the riding school. I enjoyed these visits: the boisterous camaraderie, the attention he and his friends paid me, and above all, getting out of my afternoon nap. When we went into the clubhouse, he would toss me into the air over his head before we all sat down in cushioned wicker chairs around a low table stacked with refreshments. I had my first taste of shandy—a mixture of beer and lemonade (known as 7-Up or Sprite in Canada)—on the riding school verandah.

I have fond memories of summer *siestas* spent curled up in the crook of his arm as he lay in a reclining chair on the verandah, dreamily listening to bluebottles bumping against the nearby window panes, the chatter of crickets or grasshoppers in the tall grass, and the snorts, whinnies and clumping of hoofs from the stables. The world around us slumbered under the stillness of the afternoon sun.

I can remember only one negative experience at the hands of Uncle Nicolas. He belted me once on my bare bottom for committing an offence which I cannot remember, but it was probably deserved. Of course, I ran to my mother for comfort, which she readily gave me. Curiously, I never harboured a resentment towards Nicolas for this treatment, even to this day. He was my dad, the only one, who I was just getting to know!

Stepfather, Nicolas Rojdestvin, proprietor of the Victoria Riding School in Ma Tau Wei. He was a no-nonsense military man, but we got along well for the short while we were together. Ca 1930s

Aloft or my first riding lesson. This was at the Victoria Riding School in Ma Tau Wei in Kowloon. c1937.

On one of my visits, I well remember walking nervously beside the hired hand inside the stables between two rows of stalls, keeping well clear of the snorting noses and swishing tails, and giving wide berth to the "kickers"; they were identified by red ribbons tied to their tails. Nervous or not, it was a thrill to be hoisted onto a horse and be led around the paddock by the hired hand or Uncle Nicolas, my hands sweating from holding so tightly to the crop and reins. The sheer size of the animals awed me, as did the long drop from the saddle to the ground.

§

It was in late 1935 or early 1936 when Mother received a significant commission. It was to create stage murals on canvas backdrops for the new auditorium at the Italian Convent School on Caine Road, where my sisters were in residence as boarders. Of course, during this work, Mother had to be away from home all day. At this time, we had two servants employed, Ah Kai and Ah Ngan, so I was well-cared for when Mother was not at home.

This commission was welcome but arduous work, and the lighting was not very good. It required climbing scaffolding and ladders and walking on various levels of staging all day long. There was a deadline to meet, and a lot of pressure was placed on Mother to complete the work on time.

When the opening day finally arrived, the Jubilee Concert was attended by his Excellency the Governor of Hong Kong, Sir Andrew Caldecott, together with distinguished guests. Included in a press report of March 26, 1936, was that the scenery for the different sets of the play was skillfully painted by Mrs. Tatz specially for this performance.

Trips with Mother

During the time Mother was working on this commission, she occasionally took me along to the Canossian Convent School. The journey to Hong Kong Island started by boarding a bus near our home. Children under age six rode free. Chinese passengers were usually confined to the back section of the bus, where the fare was cheaper. I loved bus rides, and always wanted to be seated right in the front, closest to the driver to watch him at work steering and especially changing gears. The smell of the diesel fuel was intoxicating!

The area around the Kowloon Ferry Terminal was usually very busy: there was the main train station and the main bus terminal; not far away were the Kowloon Wharves, which were frequently occupied by ocean liners. Stepping off the bus, we would have to walk past a row of rickshaws waiting for ferry passengers. The surge of crowds going hither and yonder added to the general bustle, as we worked our way towards our ferry, which had just disgorged its load of passengers. At certain times, the congestion of people was totally wild, and there was definitely no room for dawdlers.

The upper deck of the ferry was occupied mostly by white passengers; the Chinese thronged on the unprotected open lower deck, where the fare was cheaper. The partially enclosed upper deck commanded a great view of the surrounding scenery and all harbour activities. Under the shrill blast of the steam whistle, I watched the sailor draw up the gangway, as the ferry casted off from its berth. This was immediately followed by a violent vibration from the propeller as the seas boiled behind the stern. I was fascinated by the churning water and often wondered how fish could survive such severe turbulence. As the ferry finally got underway, the unfolding scenery immediately captivated my attention.

The scenery was spectacular with sampans bobbing and nodding in the harbour; junks in full sail, some of which were rather tattered; wallah-wallahs (water taxis) with throbbing engines putt-putting past; motor launches of all types chugging along; sometimes even a beautiful white yacht would glide along with sails fully unfurled; all types of cargo ships tied up to buoys in mid-harbour; seagulls screeching around the stern of the ferry; puffing tugboats towing barges; and the most exciting event was when we would be passing another ferry going in the opposite direction and our ferry would rock in its wake. Wow, how intoxicating that was!

Arriving at the ferry terminal at Hong Kong Central was the beginning of another adventure. Sometimes, rolling in the swell of a wave, the ferry would have to wait for the departing ferry to get underway. Finally, as our ferry began to manoeuvre alongside, the sailors threw out a tie rope to the sailor on the pier; sometimes they would miss and would have to try over again. By then, passengers crowded near the exit ramps, and the ferry would start to take a pronounced list to that side. I was fascinated to see the rubber-tire fenders crushed as the ferry was safely secured alongside the pier. Then the all-clear whistle would

The same view as crossing the harbour with mother. Destination the Italian Convent. *Photo from a street vendor. Source unknown.* c1930s

Entering HSBC. The bronze lions guarding the main entrance were named Stephen and Stitt, in honour of two former Shanghai managers. It was quite normal for children to clamber on top of these lions, and they had lots of fun. 1935.

gwulo.com
Courtesy: American Geographical Society Library, University of Wisconsin-Milwaukee Libraries.

The ferry ride back to Kowloon was a pleasure. I could see the Peninsula Hotel, and in the distance the Kowloon Park. I was so looking forward to telling Ah Kai all about today's activities with mother. c1930s

gwulo.com

sound, down would drop the two ramps and surging humanity would spill out of the ferry, making a headlong rush onto the pier and exit into Connaught Road.

That was when I had to clutch my mother's hand ever so tightly as we were jostled along! Waiting for disembarking passengers on the other side of the road were the usual rows of sedan chairs and rickshaws, all lined up for the convenience of passengers. Depending on the weather, Mother would take a sedan chair or a rickshaw, or even walk to the nearby Hongkong and Shanghai Bank, where she had an account. I didn't mind the rickshaw ride, with me usually sitting on Mother's lap, but I wasn't too happy on the sedan chair; the swaying up and down motion wasn't entirely agreeable to my stomach, and I always felt relieved when it was over, albeit somewhat a little "greenish around the gills"!

I loved clambering around the two enormous bronze lions at the entrance to the bank. In the vestibule were the elevators to reach offices on the upper floors. The main hall inside was grand and featured some enormous murals. Mother would sometimes go to her safety deposit box, and other times would be waited on by one of the tellers in the main hall.

Leaving the bank through the south exit, Mother and I walked along Queens Road, crossed over to Garden Road, then onto Ice House Street. After that, I remember climbing up shallow granite steps to mid-levels ending up on Caine Road, very close to the convent entrance. The boarding school was in the west wing of the convent grounds.

It wasn't until many years later that I learned that the school was known as the Italian Convent School. The other names it was known by were the Canossian Convent or the Sacred Heart School. For some reason, the entrance always looked rather forbidding to me, with its iron bars on the windows and front door reminiscent of the entrance into a prison.

I was intrigued by the habit the Canossian nuns wore. The head gear consisted of a black bonnet with five cylinders, approximately one and a half to two inches in diameter on top of the bonnet. On top of this cluster of cylinders was attached a black bow. A black cape covered the top half of the body from shoulder to midriff, under which was a full-length brown bodice reaching to the ground. The most intriguing part of their habit was the head gear.

I remember meeting some of the nuns, and very probably Mother Gina, under whom Mother was working. I don't remember seeing either of my sisters; they were probably attending classes at the time of our visits. I also don't remember what kept me occupied during the day while Mother was at work. Very likely the nuns kept me busy, but how is totally outside my memory.

The return trip back to Kowloon and home varied from time to time. The diversions sometime included a walk down from the convent toward Queen's Road, and perhaps a quick stop at Central Market on Des Voeux Road. A stop at the Sincere or Wing On department stores was of great interest to me, especially when Mother would buy me a treat. I was always fascinated by all the ready-to-hand merchandise so readily available in these department stores. Of course, going up the elevators in a cage with an operator at the controls was so intriguing, as I watched the moving scenery below me.

Riding on the top deck of the double-decker trams was always an exciting experience, especially as it swayed along on the tracks. I loved riding on the top deck, with my mother in tow as we clambered up the narrow stairs. I wouldn't dare stick my head out the window when passing trams on the opposite track and would hang on for dear life as the tram would take a sharp turn with wheels screeching. With two trams approaching each other, I often used to duck in anticipation of an inevitable collision and was relieved when we passed each other without incident. A feature about the trams that always caught my attention was the ability of the seat backs to be swung from front to back depending on the direction of travel; and the operator having to move from one end of the tram to the other end at each terminal.

Walking along the Praya, known as Connaught Road, towards the ferry terminal always made me feel nervous; there were no protective railings along the water's edge. Numerous boats were tethered to the retaining wall, and to board the boats it was necessary to literally "walk the plank", something I would never attempt. There was not as much bustle around the area of the ferry terminal on the Hong Kong side, so I felt more at ease with my surroundings. Getting onto the ferry, I would hurriedly find a vantage point and press myself against the railings. Once again, the changing view would grab my attention; maybe this time there were different boats coming into view. Again, I was fascinated by the rocking of the ferry as it got "washed" by passing boats. Disembarking on the Kowloon side, Mother and I would get on a bus to take us back home in Ho Man Tin. Of course, it was a welcome sight to see Ah Kai, who would give me her attention, and also help Mother with whatever parcels she might be carrying from shopping.

§

From May 2 to June 3, 1936, Mother was away from home in Canton (Shameen) on a contract to provide art painting lessons to some students. This information is contained in her application for British naturalization, dated June 29, 1938.

Father had been a recognized art instructor from his experience in Harbin, and this reputation probably followed him when the family moved to Tientsin, where he was well-known in the local art circle, and finally to Shameen in Canton. By this time, Mother's own art skills must have matured significantly.

An important patron of my parents was Mr. Weirink, the Dutch consul for Canton and Hong Kong, who had become a close friend of the family. They were living at the Repulse Bay Hotel on Hong Kong Island. No doubt this connection was of benefit to both my father and mother.

In Care of Ah Kai, and Our Outings

With Mother absent for a significant period and Stepfather busy at work during the day, Ah Kai had an important role in caring for me as a preschooler back home in Ho Man Tin. My memories of Ah Kai are strongest from the time that Nicolas first appeared in my life, but first I should explain more about the role Chinese domestics played in the life of expatriates in Hong Kong.

Chinese domestics are sourced by reference, usually from a domestic already employed in another expatriate household. The domestic refers to the head of the house as "Master" and the wife as "Missy". The domestic always reports to the missy for instructions on a daily basis. All domestics, including Ah Kai, wore a white smock with a high collar that had Chinese-style buttons on the side, complete with black pants. Ah-Kai also wore soft-soled, cotton-top black shoes around the house. To go out, she would change to another outfit and often wore wooden clogs. At times, she wore her hair in a braid, but often she would coil it into a bun on top of her head with a spike through it. That is how I remember her best.

The duty of the domestic is to run the household under the supervision of Missy. This includes daily shopping in the market, cooking and serving of meals at the table and cleaning of utensils and crockery. They are also responsible for cleaning the house, including laundry of clothing and linen, and subsequent ironing.

When parents are unavailable, full care for preschool children has top priority, which may include dressing them, feeding them, bathing them, amusing them at home or in a nearby park and at night getting them to bed. The domestic is in fact practically a member of the household and, because of this, is provided with quarters on the premises.

Diana Fortescue wrote in *The Survivors*: "*Domestic life was in the hands of servants who were also part of a super-efficient intelligence network. At dinner parties, if the host had insufficient place settings for a special occasion, the servants would go to the neighbours to borrow whatever was lacking (with the aid of their servants, but without the owners' knowledge). It was not uncommon to go to a neighbour's dinner party and see some of your own china, glass or even silverware on the table.*"

If domestics were overwhelmed by work, they usually brought in a family member to help. If the expatriate family was large, then of course the missy would engage more domestic help on a permanent basis. In Ah Kai's case, her backup was Ah Ngan.

Ah Kai had all the above responsibilities, and it wasn't too onerous as she had only Nicolas, Mother and me to look after. When my sisters were home from the boarding school for the weekends and school holidays, Ah Kai used to bring in Ah Ngan to help with the laundry and to escort the children to a nearby park. Margaret was old enough to help mind her siblings from time to time, but I have no memory of any participation from her.

All this enabled Nicolas and Mother great flexibility to enjoy a good life together when they were not at work. Under these circumstances, it is not surprising that Ah Kai and I developed a strong bond through intimate contact with each other. However, it in no way diminished parental influence and affection, because daily contact was still maintained by Mother.

For practical reasons, based on trust, Ah Kai was permitted to take me out with her on some of her market excursions. Ah Kai did more than that when she took me out to visit her own family members.

Mother must have known about this from conversations with me, but as far as I know, she never took Ah Kai to task.

Looking at Mother, I could tell that she was very happy. How did I know? Because I started to see less and less of her around home; she and Nicolas were very busy going out in their new social life, as well as both being occupied with their own activities. Ah Kai handily stepped into my life and dedicated herself to caring for me when Mother was unavailable, and in this period, Mother was often away from home.

Growing up in Hong Kong from a young child's perspective was captivating, with intimate, albeit brief, exposure to a kaleidoscope of Oriental culture. Instinctively, young children tend to observe life through rose-tinted glasses, which is enhanced by the presence of trustworthy caregivers. Memorable childhood adventures were gained from exposure to the indigenous population, where segregation raised questions from time to time, but were inconsequential in the eyes of a young child. Fortunately, with limited experience, young children soon accept what comes to them as universally established.

§

"*Ah Kai, nay hui binsui ah* (Where are you going)? *Tung ngor ah* (Wait for me)!" I called out to her one morning as she was getting ready to go to the market in the Old Kowloon Walled City. Ah Kai patiently bent down to buckle my sandals, while I sprawled onto her back, hugging her. I found great comfort in the intimate warmth radiating from her body.

It was always fun to go out with Ah Kai. Together we would walk towards the market in the Old Kowloon Walled City, and as usual, push our way through crowds at one of the gateways. Invariably, the market was teeming with a mass of people engaged in all kinds of activities, as vendors and shoppers could be heard haggling at the top of their voices. I was no stranger to the market, so I was quite at ease with my surroundings, especially because I enjoyed a good command of the Cantonese language.

I was fascinated by the live eels swimming in wooden tubs and watched them squirm, as with my fingertips I tentatively touched their slippery scales; I also looked at live Garoupa (a common tropical fish) swimming in aquarium tanks or wooden buckets. It was not unusual to see salted minnows laid out on newspapers on the sidewalk, drying in the sun, samples of which I would love to chew on. I felt so sorry for the live chickens hanging trussed upside down by their legs in readiness for

Thewellman, Wikipedia commons

It was not uncommon to find a houseboat village outside the confines of a typhoon shelter, such as the above. It was in one of these that Ah Kai had other relatives whom we visited. What appealed to me was the closeness of family members in a cozy environment. 1970

Courtesy Mike Cussans

One of a few little villages in Kowloon. Note the happy toddler and free range chickens. Life in these villages always intrigued me, and I felt quite comfortable with the locals, thanks to Ah Kai. c1960

Courtesy Mike Cussans

The sampan provides a handy means of transportation within a typhoon shelter. I used to be very apprehensive in clambering aboard such a vessel, but with the help of Ah Kai, all would go well. c1960s.

gwulo.com

Countryside children – big sister caring for siblings. c1930s

Courtesy Mike Cussans

They are ethnic Hakka Han Chinese, originally from the north, but had migrated to the southern provinces where they became known as Hakka. They have their own Chinese dialect. c1960s a buyer. There were heaps of fresh vegetables of all sorts and an exotic array of fruit on display. Then there were the 100-year-old salted duck eggs encased in charcoal. Stalls of cooked meats always had such an enticing aroma. It was all so fascinating, with never-ending surprises at each visit to the market.

As usual, the atmosphere within the market was quite festive, with throngs of people jostling and pushing their way through the narrow aisles, with me trudging along behind Ah Kai. The vendors were highly amused at seeing the little blond *gwujai* (devil boy) chatting away in Cantonese.

Using a crude but functional balance bar, I watched as produce was weighed. The measurement of catty is equal to 1.3 pounds. Energetic bargaining would take place with the shrill voice of the vendors extolling the virtues of their goods, and the buyers haggling the price until they settled on an amount, everybody smiling from ear to ear and all satisfied that they had the better part of the bargain.

We never ventured too far into the walled city but did see the odd shops on our way to the exit: a busy cobbler at work, a weaver of rattan baskets and even the odd herbalist's shop. The three or four communal water taps on the ground floor were always in great demand for laundry and domestic purposes by the residents of the surrounding tenement buildings.

From time to time, there would appear jugglers performing on the street sidewalk. Elsewhere in the colony, there were restrictions for this kind of activity, and the police would often interfere when they came around, but they seldom bothered within the walled city, as their jurisdiction there was limited.

The Chinese love a crowd, and the first indication of something interesting happening was of course to see an unusual gathering of people. Pushing our way through them, at one time I was spellbound watching an acrobatic performance; it was an older male accompanied by a young girl, who appeared to be seven or eight years old. With mouth agape, I watched her, under the direction of the man, who was obviously her trainer, go through incredible acts: she would go through her routine walking on a pole with her bare feet, tumbling with great agility, performing incredible feats of balance and even bearing the full weight of her director in some contortionist acts. It was all so mesmerizing that Ah Kai had to drag me away.

§

Ah Kai, bless her heart, always took pride in showing me off to her friends and family members, sort of as her protégé. There wasn't any extensive Chinese conversation between us, just exchanges of useful phrases, which I quickly learned. What kind of conversation could I contribute in any case with limited experience at that age?

Another occasion was an excursion to the countryside. Ah Kai announced to me, "*Gum yut, ngor dei hui tai ngor ga garjiar* (Today, we will go to see my sister)." She usually employed the Cantonese vernacular with me, with which I was quite familiar. I have no recollection how we got to our final destination in the countryside, but I remember well a few distractions along the way.

An unavoidable and common feature in the nearby countryside was the presence of numerous cesspits or dugouts, strategically located nearby vegetable plots, and amongst some of the smaller paddy fields. "Night soil" was regularly gathered from nearby housing and farm dwellings in portable "honey" pots, which were then emptied into these cesspits, ultimately to be used as fertilizer. As needed, the farmer would approach one of these cesspits with two wooden buckets balanced on a single pole across his shoulders. These wooden buckets would be submerged in the muck, and when full the farmer would stagger out, and through a spray nozzle would methodically sprinkle the nearby vegetable crops. Unsanitary as it was, it was still a widespread practice and an efficient method for disposal of raw sewage which produced startling results in robust vegetables. Imported vegetables were preferred from establishments like Lane Crawford, where expatriates shopped.

As could be imagined, the odour was overpowering, and the occasional whiff made me gag. A good reason why the white population would never knowingly use local vegetable produce without scrupulously washing them in permanganate solution, a dark purple crystalline disinfectant, followed by thorough cooking. At times, permanganate grains were also used for bath water, and it was always fascinating to observe the dissolving grains gradually transforming the bath water to a pink colour.

As we carefully picked our way towards our destination, I found it captivating, nestled in a grove of trees, to observe farmers wearing peaked straw hats, busily ploughing their muddy paddy fields. I felt envious to see the farmers squishing their bare feet in the mud, walking behind rustic wooden ploughs harnessed to black water buffalo. Ah Kai would have to urge me to move along from time to time. At the end of the trek, we came up to a small cluster of shacks and buildings near which were some shade trees.

As is common in China, relatives lived mostly together within their own communities, sometimes in a walled village, and other times near each other, and consequently, many of them share the same family name. This was also a method to provide support and security to each other. It was indeed a happy environment!

The reception from Ah Kai's relatives always included a lively and noisy welcome, either in the courtyard or inside the farm dwelling. There would be the older folks; some of the women minding the younger children, some still displaying the effects of bound feet; adults all chatting away and giving a lot of attention to the *gwujai* (devil boy), who at first, clung to Ah Kai.

It wouldn't be long before I would respond to the opportunity to react with a few of the Chinese children of my own age, as they would also eagerly "show me off" to some of their neighbouring friends. The appearance of a "white" human being in their surroundings was always cause for great excitement, especially in the countryside, where white people in those days seldom ventured.

Of course, there were live free-ranging chickens, some in and out of the house. The fleet-footed chicks were always so much fun to try and catch. There were also the inevitable mangy-looking chow dogs, scarred from dog fights. It was also my introduction to witnessing the practical way Chinese parents addressed babies' or toddlers' need to relieve themselves, naturally and without the use of diapers. The toddler's pants had a convenient slit through which the infant or toddler would function nonchalantly as needed. It was all in keeping with nature, and what happened afterwards I never wanted to see!

§

On another occasion, Ah Kai and I also visited another family group living on a houseboat in the Typhoon Shelter at Yaumati: they were known as Tanka People. As I was terrified of deep water, Ah Kai had to carry me on her back to get onto the bobbing water taxi, which ferried us to the houseboat. I was amazed to see a young girl nonchalantly navigating around the Ya Ma Tei Shelter with one oar, even with a baby strapped to her back.

The houseboat was large enough to provide accommodation for an entire family, including a dog or two. We clambered aboard to a great reception from Ah Kai's relatives. I think the visit was short at the time, so I have little memory of any significant activities. However, I do remember lying on my stomach on the deck and peering over the edge and into the water below, watching marine life. What I saw were swarms of little minnows nibbling at the growth of seaweed attached to the hull. Occasionally, a large fish would dash amongst the minnows and scatter them in all directions. It was also interesting to see how the sunlight penetrated into the murky water. Then, Ah Kai called me to get ready to leave.

§

As indicated earlier, Ah Ngan appeared in the household from time to time to give assistance to Ah Kai, either with housework or care for me and my sister Julie, when she was home from the boarding school.

On a typical outing, Ah Ngan would take us out walking in the neighbourhood and to a nearby park. In those early years, the district had numerous sections of undeveloped raw land, which provided me with wonderful opportunities to explore on my own. Lifting rocks often readily revealed scurrying water cockroaches, beetles and sometimes centipedes; earthworms too would be evident. On one outing, I vividly remember a painful experience. Unknowingly, I happened to be standing on an anthill, and without me feeling anything, tiny red ants started swarming up one of my bare legs. When I finally became aware of them, I naturally stepped away and began brushing them off; that was when I discovered that these were a vicious species of fire ant. When disturbed, these ants will sting the offender mercilessly. My screams brought Ah Ngan running, and she frantically brushed the rest of the ants off my leg. From then on, I stayed clear of fire ants and associated only with the larger black ants that have no sting in them at all.

Among the local flora were two kinds of bushes that intrigued me: a thorny bush that bore a profusion of multi-coloured tiny blossom clusters, and a bush with small fern-like leaves that would close like a trap at the slightest touch of a finger and, in time, open up again. This is the "sensitive plant", and I delighted in running around touching all the fern leaves in my reach and watching them close before my eyes.

§

It was inevitable that my close association with Ah Kai and Ah Ngan at this young age helped to develop my ability to converse in the Cantonese language. It was common for preschool children and their

Chinese *amahs* (a girl or woman employed as a domestic servant) to form strong bonds, and more than one child brought up in Hong Kong learned to speak their *amah*'s Chinese dialect, sometimes even before they learned English. Under Ah Kai's tutelage, I got to know the native Chinese people in a way that most expatriates never did. I mingled with them, ate their food, learned their customs and spoke their language.

§

One of the last significant events Ah Kai shared with us was the massive typhoon of August 1936. It pummelled Hong Kong for about 12 hours, and although it didn't do much damage to housing on land, a maximum wind force of 130 miles per hour created havoc to shipping in the harbour; 11 large ships and numerous smaller crafts were sunk or wrecked. What I can recall was the typical torrential rainfall during and at the end, and seeing the central courtyard in our apartment block flooded with water. Another incident I heard discussed later was that after the storm had passed, some of the Chinese people were electrocuted when they attempted to clear some downed power lines, not realizing that they were still energized.

A household servant, Ah Kai was with the family as long as I can remember until the move to Nathan Road. By then she had been with the family for over eight years. She was very special to me and cared for me intimately. , c. 1930s

I heard nothing about the condition of the Matshed stables housing the horses at the riding school, or the condition of the horses. Because of the flimsiness of the Matshed, there must have been some damage done there.

Death of My Step-Father, Uncle Nicolas

Sadly, Mother's new lifestyle was about to be tragically terminated, leaving her and her family totally devastated, from which we would never fully recover.

Sometime in the last week of December 1936, Nicolas took a severe tumble from a horse and was rushed to the nearby Kowloon Hospital, where he succumbed to his injuries. I have no recollection of the events surrounding the accident, or those immediately following his death. Tragically, this left Mother once again a widow. Sometime later, I remember Mother taking me to visit his gravesite at the Happy Valley Colonial Cemetery. Mindful of his passion for horses, Mother had carefully chosen a burial plot which provided a view of the racetrack at the Happy Valley Jockey Club.

Subsequent research has revealed an obituary notice in the South China Morning Post on page 12, dated December 31, 1936, stating Captain N. A. Rojdestvin, aged 42, and former cavalry officer of the Tsarist regime, died after midnight in the Kowloon Hospital. Mother was named the chief mourner at the funeral service in the cemetery, and curiously, the list of mourners included a Mrs. O. M. Robinson. I am positive that O. M. Robinson was Olga Mary Robinson, the same person who eventually became my godmother in the spring of 1939.

Mother, now 32, was widowed for the second time in five years, just when a new and promising life had seemed within reach. As for me, the impact wasn't as great. Sure, at first, I missed his presence and the frequent trips to the riding school; a riding school that I was never to step foot into again. The riding school operation was taken over by an associate of Nicolas, a former Colonel Litvinov, also of the Tsarist regime.

CHAPTER 3: ON NATHAN ROAD

Getting to Know Mother

1937 – 1939

After Nicolas passed away, Mother didn't waste any time and moved to the third (top) floor at 241 Nathan Road in early 1937. She probably discontinued her gift shop operation at Estelle Dress Shop at the same time.

Margaret and Julie, by this time, were well-established as boarders in the Italian Convent Sacred Heart School, and it would be later this year that I would be enrolled in the Primary 1 Grade in a nearby school. Mother had time to settle into the new accommodation before Julie and Margaret would be spending the summer holidays at home. Mother was spending more time at home, so Ah Kai was only called in to help on a part-time basis.

To support herself and the family, Mother had decided to open an art studio at home, from where she would execute portrait paintings in oils and offer art lessons. Once Mother was settled in the new apartment and had her studio properly set up, she became a member of the Hong Kong Art Society, and entered the following advertisements in the South China Morning Post (SCMP) newspaper:

> *August 3, 1937 Page 5 ART STUDIO*
>
> *Mrs. A. Tatz gives lessons in painting, drawing and sketching portrait, figure, still life, etc. Also classes for children. 241 Nathan Road, top floor.*
>
> *October 31, 1937 Page 4 TUITION GIVEN*
>
> *Mrs. Tatz gives lessons in painting, drawing, sketching, etc. (portrait, figure, still life, etc.). Also accepts orders for portraits and advertising sketches. 241 Nathan Road.*

I never did see or know how successful she was in acquiring students, but by the pictures on the wall of the studio, she appeared to specialize in portrait paintings. She must have been very active during the day after I started school in September 1937. I only remember one sitting, when one winter evening she brought a blind old street beggar into the studio for several sessions. She paid him and fed him at the end of each session. This painting is still in the family.

There is evidence to suggest that Mother's reputation as an artist was gaining renown in the art circle of Hong Kong. The following were sourced from South China Morning Post:

> *March 28, 1938 Page 18, Critic Reviews All Exhibitions During 1937, And Looks Ahead.*
>
> *The Hong Kong Working Artists Guild will hold a reception next month at No. 2 Connaught Road Central, in honour of His Excellency, the Governor's visit to the Guild, during which a collection of representative paintings by local artists will be displayed. We publish below a survey of the Art Exhibitions held by the Guild during 1937, contributed by the Guild's Vice-President, Mr. Luis Chan.*
>
> *The following was mentioned under "Art for Relief": A charming self-portrait executed in oil by Mrs. Tatz.*
>
> *April 9, 1938 Page 9, Artists at Home exhibition held in the studio of Mr. John Locke at 2 Connaught Road, attended by His Excellency the Governor, and supported by local artists. Also represented were the Travel Association and the Education Department. Artists who sent in work for the At Home included Mrs. A Tatz.*

Here, usually after supper, Mother and I spent a lot of quiet time together in the evenings. Mother would clear away the evening meal and proceed to work in her studio, which was really an extension of the living room. I would watch her prepare her material, lay out the tubes of paint, get her paint brushes organized and get set up for the night's work. I didn't know what it was, but it was so comforting to feel such a closeness to Mother, just the two of us together, with my dog Rexie.

After a while, I would turn to my wooden toy soldiers in the living room and march them along the border of the carpet. Some soldiers were on horseback and others on foot, and they all guarded a fortress of interlocking wooden pieces. Many a battle was fought on that carpet, and England won them all, because I was the British general leading a British army. On other occasions, I would drive my toy cars up and down ramps until I would hear Mother suggesting that I should start getting ready for bedtime.

§

When my sisters were home from the boarding school, Mother occasionally sent me and Julie off, with my oldest sister Margaret in charge, to the cinema. The Majestic and the Alhambra were both within walking distance from our flat, and we sometimes also went to the Star Theatre on Peking Road. The order of entertainment from the cinema was always the same: previews, newsreel, cartoons and a short intermission before start of the main feature.

The Alhambra Theatre had a huge, round, ornate chandelier hanging from the ceiling that dimmed as the time ticked down to the start of the show. By then, the three of us were seated and enjoying our treats. At the short intermission break, we joined the crush to replenish our treats at the confectionery and quickly hurried back to our seats before the cinema was plunged into darkness for the main show. A few times, I had to grope my way back in complete darkness to find my seat.

At times, when nature called, I was very reluctant to tear myself away from the screen (unless it was an uninteresting newsreel). On those occasions, I would be chastised by my sister Margaret, as she would say, "I told you so!" to go along with my damp trousers.

For days after each show, I used to re-enact various scenes by myself, or with anybody else who might be around. I could be pretending that I was a cowboy in the person of Gary Cooper, or the swashbuckling Tyrone Power in the Mark of Zorro. This world of fantasy was very real to me, as I also engaged in mock battles with Indian war tribes, and particularly battle scenes I was familiar with from Rudyard Kipling stories. All the sounds of battle were issuing from my mouth in a constant stream.

When I was called to table by Mother for dinner, I had first to tie up my "horse in the stable and put my weapons safely away." Then I would have to wash my hands before coming to the table. Mother might ask me, "How many bandits did you shoot today?" or "Who is Hoppalong Cassidy?" With a condescending sigh, I would patiently explain to her what was what! Mother, being who she was, was always very understanding.

Nathan Road facing north. Left arrow pointing to top floor of 241, our home where we lived from 1937 to 1939. The right arrow is pointing to the Kriloffs flat. This image hasn't changed much between 1939 and when it was taken, except for the newer model of buses. c1950s

gwulo.com

Diocesan Girls' School, image from a postcard. This was where I enrolled in Grade 1 in 1937 when living on Nathan Road with mother. This would be the school that I would be boarding at after mother's death from September 1939 to December 1941. Note the gate! c1930s

Robbie, Ah Ngan, Mother and Rexie, c1938

Nathan Road was a popular venue for many processions, and our top-floor apartment enjoyed a commanding view of the street from the verandah. Weddings and funerals were most frequent, and processions were always preceded (and sometimes followed) by marching musical bands.

Firecrackers are traditionally used to frighten away evil spirits. They are usually assembled into long strings, hung from long bamboo poles and connected by a continuous fuse. The intensity of the explosions varies and usually the loudest is the last one to ignite. Firecrackers are usually discharged at the end of a procession, and the explosions create a shower of red fragments, an auspicious colour to protect the dead in a funeral or to honour a bridal couple.

I used to watch these proceedings by hanging on the verandah balustrades, with either Ah Ngan, Ah Kai or even Mother hanging onto the rear of my pants so that I wouldn't fall over into the street.

§

The move to Nathan Road coincided well with my entry to Primary Level 1, as I would turn six years old in another month. Not only that, but the school was conveniently located at the junction of Gascoigne Road and Jordan Road, an easy walking distance from where we were living on Nathan Road. The Diocesan Girls' School (DGS) was founded in 1860 by the Anglican Church.

Although DGS was classified as a girls' school, young boys were accepted in the primary grades, and they were not required to wear any uniform. It was expected that these boys would eventually transfer to the Diocesan Boys' School (DBS), where they would complete their formal education. Both schools offered boarding facilities.

The highlight of the first day was me wearing a blue dungaree (it wasn't made with denim), which Mother had made for me from a pattern in the Montgomery Ward catalogue. Main memories from attending school were listening to Enid Blyton stories read by our teacher, Miss Crawshaw; my role as the ugly duckling in a school play; and a confrontation with a class bully who was after my lunch. Apart from that, I think I survived quite well.

For Class 1 students, classroom work was not demanding. We practised memorizing the alphabet by singing the letters out loud. Script copywriting, I found to be quite tedious, but it was a requirement to develop proper writing skills. After several pages of this exercise, my second and third fingers on my right hand started to ache from

Group picture on rooftop of 241 Nathan Road, view to the east; Julie, Margaret, Robbie, Mother. c1938

Last group family photo, taken in Mother's studio on Nathan Road, winter 1938/39.

tension. We only used wood lead pencils. As for sums, I didn't find that too bad, as far as addition and subtraction were concerned.

On September 2, 1937, an unnamed typhoon was one of the worst in the history of Hong Kong, killing 11,000 people. It raged for only about seven hours with 164 mile per hour winds, with gusts estimated at over 200 miles per hour. Twenty-six large ocean-going ships, including two of 20,000 tons, were either sunk or wrecked.

Mother and I had barricaded ourselves with wooden shutters on all the windows. But on the top floor, the howling wind and flying debris beating against the shutters was intense, like the wrath of demons. The shutters all held fast, and when it was over, Mother, I and Rexie, our dog, breathed a great sigh of relief. Of course, debris had littered all the nearby roads profusely, which in time would all be cleared up.

§

I recall very little about Christmas and Easter celebrations during this period of my life. I never really understood the significance of these events, and thought that Santa Claus represented Christmas Day, and the symbol of Christmas was a special tree all lit up with candles; I remember nothing about the Nativity story. Easter was identified by coloured hard-boiled eggs, a Russian tradition, and a dreary Russian Orthodox church service, listening to sonorous singing of hymns.

I vaguely remember discovering a toy car in a box meant for Christmas (the one that Julie wrecked before my eyes) when living in Ho Man Tin. When Mother found out, she quickly retrieved the car, and by 'magic' it reappeared again at Christmas time.

Christmas trees I remember were of the Jack pine variety, festooned with decorations and real wax candles clipped to extended branches. Candles were only lit in the presence of adults and never left unattended, due to the extreme risk of fire. I vaguely remember a Christmas tree in December 1937 in Ho Man Tin, and don't remember anything in Nathan Road in 1938.

We were living in Nathan Road in 1938, when Mother allowed me to help her colour Easter eggs. Different coloured dye solutions had their own pots, and when all colouring was complete, Mother allowed me to fool around with the solutions. I thought I would mix all the solutions in one bowl and was completely surprised when my egg came out all black instead of multi-coloured. Mother chuckled at my surprise. This would be the last Easter I would have with her; the next one would see her hospitalized.

Mother's Friends

As far as I am aware, my parents did not have many friends living in Hong Kong when they first arrived. At the beginning, there may only have been a Mr. and Mrs. Weirink, the Dutch consul and his wife. They were good friends of the family from their time in Canton (Shameen). The following are my own reflections, but Mother might have had other friends that I was unaware of or too young to remember.

Outstanding friends to Mother and our family were Costia and Lydia Kriloff. They were almost our neighbours on Nathan Road at the northeast intersection of Jordan Road, and because of that, we often visited with each other. They too were Russian refugees from North China.

Constantine Alexander Kriloff was born in Russia on January 6, 1905. He had enlisted in the Hong Kong Police Force on May 27, 1930, as L/Sgt. E6. He was a member of the first Anti-Piracy Guard assigned to the passenger ship *Empress of Canada* on its way from Hong Kong to Shanghai. By the time he became our friend, he was working in Special Branch, Passport Enquiry Section.

Uncle Costia and Aunty Lydia Kriloff lived in a flat opposite ours on Nathan Road. Mother often visited them and we all became good friends. I also got to know Vitaly Veriga and his first wife, as well as his mother; Vitaly was Lydia's brother. Both Costia and Vitaly were members of the Hong Kong Police Force. I do not remember ever seeing Uncle Costia in a policeman's uniform, but he always went to work armed with a revolver in a special holster under his jacket. He was a lot of fun, and I always looked forward to visiting him and Lydia.

Without realizing it at the time, 1938 would be my last birthday celebration, not only with Mother and friends, but for the next 14 years. My seventh birthday was special, even though I have no memory of what gifts I received from Mother.

What was so special about it? Two reasons: firstly, the party was at the home of Uncle Costia and Aunty Lydia, and secondly, the amazing cake! It was a huge chocolate-coated vanilla ice cream cake with seven lit candles. I gave a big hug to Mother, Uncle Costia and Aunty Lydia to express my happiness.

I feel quite certain that it was Costia Kriloff who advised Mother to apply for British naturalization. References shown on Mother's application, dated June 29, 1938, were Mr. George William Arnold, Mr.

James Downie Thomson, Mr. Hubert Gladstone Williams and Mr. Francis John Thomas Locke.

The application was approved, and Mother received her British Certificate of Naturalization, effective January 14, 1939. This certificate was signed by the Governor of Hong Kong (Northcote) and approved by the Undersecretary of State. The final procedure was the Oath of Allegiance to King George Sixth, carried out on January 31, 1939, by the Deputy Registrar of the Supreme Court at the Courts of Justice, Victoria, Hong Kong.

§

Like Mother, Mr. and Mrs. Nesteroff and their families were also victims of the Russian Revolution and Civil War. They lived on Nathan Road (just south of Austin Road), in an apartment on the second floor of a building near Cherikoff's Bakery. Their two sons, Yuri and Igor, were a few years older than me. Mr. and Mrs. Affanasieff, parents of Mrs. Nesteroff, also lived with them. Mother often took me to visit this family, and in return, the two boys used to visit us, and I had fun playing with them.

Olga Robinson and her mother, Catherine Dobrjitsky may have lived somewhere in Hung Hom, but later moved to Taipo Road. They were of Polish origin and were Roman Catholic. They will appear again and became important later on.

Mrs. Kovac and her teenage daughter, Billie, were Austrian and could originally have been friends of my father. I think Billie was a student at the Sacred Heart School with my sisters. It was many years later that I remember meeting Mrs. Kovac at Mother's gravesite in Happy Valley.

Mrs. Karpusheff, who owned the Estelle Dress Shop on Mody Road in Kowloon, was an early friend of Mother. She and her husband, Boris, might well have been Russian refugees from North China. Mrs. Karpusheff was always warm towards me and gave me attention whenever I visited her shop with Mother.

Mr. and Mrs. Rodgers, an American stockbroker and his wife. They were living on the Peak, possibly on Coombe Road. Mrs. Rodgers was Russian and might have originally been a refugee from North China. I do not remember Mrs. Rodgers (Elizabeth) visiting Mother, but after Mother had died, I was surprised when she brought me to her house for a Sunday brunch. Subsequently, Mr. Rodgers was killed in battle in the Defence of Hong Kong.

Henry Corra and Poldi, his wife. From available family records, their granddaughter, Susan Ponsford, feels certain that somewhere in China, very likely in Tientsin, our two families were acquainted. Susan's mother was Christine Genders (née Corra). I first met Christine at one of Elizabeth Rodger's Sunday brunches before the war. Our two families were Austrian nationals at the time and we might have been neighbours living in one of the International Concessions in Tientsin (Tianjin). According to Susan Ponsford, Christine Corra (Susan's mother) was enrolled at the Italian Convent Sacred Heart School on Caine Road while my sisters were in residence there. Did my mother and Christine's parents socialize at the school's Jubilee celebrations in the Spring of 1936? Intriguing question.

The Tkachenko family ran a bakery and restaurant in Kowloon on Hankow Road, and they knew Mother very well. Mother had developed a strong friendship with one of their adult daughters. They too were Russian refugees from North China.

Left:
Display of painted portraits in various media in Mother's studio. I recognize many of them, especially the one of Lydia Kriloff with a white collar. c.1937

Mother's last portrait, made for her citizenship application and British passport in 1939.

CHAPTER 4: PERIOD OF LIMBO

You're a Big Boy Now

1939

As often happened, after putting me to bed for the night, Mother would continue working in the studio. One night in early 1939, I was woken up from sleep by the sounds of Mother's friends who had appeared in the apartment, seemingly quite agitated. I was surprised by all this unexpected commotion, not realizing the seriousness of this event. Mother had collapsed in her studio from some kind of seizure.

After Mother was taken to the Matilda Hospital, I was initially billeted with the Nesteroffs, which delighted me as I was looking forward to playing with their boys, Yuri and Igor. The whole family was so nice to me, and as unusual as the situation was, I thought this to be a sleepover party.

At seven years old, nobody wanted to put fear into me, and nobody had any idea of the gravity of Mother's condition at the time. That night, and probably for the next day or two, I remained with the Nesteroffs in their home, while Mother's friends were trying to come to a decision for my immediate care. This was how I eventually found myself moved to Aunty Olga's home, who was almost a total stranger to me; I couldn't even remember who she was in our family's life.

In Care of Aunty Olga

I felt apprehensive when I was delivered to Aunty Olga's house, but I must have received sufficient assurances that my dislocation was only temporary. Nobody knew at the time that Mother wouldn't survive her illness.

I was appalled when Aunty Olga told me in no uncertain terms that she disliked children. In her defence, she herself was an only child, and at 37 years old, she had been widowed or divorced, had no children of her own and was in a romantic relationship with Harry Blake. With no experience in raising children, she must have been reluctant to take on the responsibility of my care, and possibly even resented my intrusion into her life.

Worse than that though, was her declaration that, since I was now a big boy (at seven years old), I should be known as Robert rather than my family nickname, Robbie. The finality of it left me stunned. I withdrew into myself and spent several anxious days wishing my mother would hurry up and fetch me home.

Aunty Olga's home on Taipo Road was too far for a seven-year-old to walk to DGS, so I was kept out of school for the rest of the school year. This was the first significant interruption of my formal education.

§

Olga Robinson was born on February 5, 1902, to a Polish father and a Ukrainian mother, Casimir and Catherine Dobrjitsky. After her father died, she and her mother fled the Russian revolution and civil war and moved to North China. She subsequently married Mr. Robinson when they were living in Shanghai, and in Hong Kong, her language skills enhanced her employment opportunities. Besides being fluent in French, she also spoke Russian, Ukrainian, Polish, English and a little Japanese, from having worked at a Japanese bank in Shanghai. In Hong Kong, she was employed by a French-speaking Belgian firm.

Aunty Olga was caring for her aging mother, who was living with her. I was told to call her mother Babushka, though since she spoke only Ukrainian, and I spoke only English and Cantonese, we had difficulty communicating effectively. Olga and her mother were living on the second floor at 105 Taipo Road, Kowloon. A significant other person in her life at this time was Harry Blake, who was her constant companion and who was often visiting at her home.

Initiation into the Catholic Faith

I do not know if my parents were wedded in a Catholic or Russian Orthodox church in Harbin; my father was Catholic, and my mother was Russian Orthodox. The Catholic religion requires that children born of mixed marriages must be brought up Catholic. Probably in response to my mother's request made from her hospital bed, Aunty Olga undertook to ensure that I become indoctrinated in the Catholic faith, and this is how she became my godmother.

Aunty Olga was friendly with several secular Polish Catholic priests who lived across from her home on Taipo Road. So, when I was living with Aunty Olga, it was there that I commenced my initial indoctrination in the Roman Catholic faith. I cannot remember being enthusiastic about this experience, especially since I had no choice in the matter, but endure it, I did.

With Mother, my experience in attending church was very limited. I recall some exposure to the Russian Orthodox services, probably at Easter, but I was far more familiar with the practice of Buddhism because of my relationship with Ah Kai, my caregiver at my most impressionable age.

The concept of God and the Holy Trinity at seven years old was beyond my comprehension. Who was this important person called Jesus? Is he God? How could anybody kill him if he created the world, I wondered? I didn't dwell too long on a subject that had little or no meaning to me at the time. As far as Buddhism was concerned, I found it difficult to understand how some deities could have such fearsome images, and at the same time others were so congenial—like the image of the happy Buddha! I knew there were evil spirits; after all, weren't exploding firecrackers in the Chinese custom meant to scare them away at all auspicious occasions?

I endured several weeks of intense indoctrination, including memorizing the words to three important prayers and learning the meanings of those words. The hardest one for me to commit to memory was the Apostles' Creed. Knowing the Creed was a prerequisite to the Confirmation ceremony, and this also meant preparation for Reconciliation (Confession). This was a lot to take in all at once and quickly. Any seven-year-old faced with a barrage of foreign, confusing and overwhelming concepts and information could be forgiven learning slowly and without enthusiasm, as I did at the time.

About six weeks later, on Easter Saturday, April 8, 1939, I was baptized by Fr. Rossi at the Rosary Church on Chatham Road in Kowloon and accepted into the Roman Catholic Church. Two days later, I received my First Holy Communion and Confirmation, administered by Bishop Msg. Valtorta, also at the Rosary Church. My sponsors were recorded as Stanislaus Adam Semenuck (probably the Polish priest instructor) and Olga Margaret Mary Robinson. Aunty Olga was now officially my godmother. The only void in this entire proceeding was the absence of my mother, who at the time was languishing in the Matilda hospital.

Life with Godmother, Harry and Babushka

Godmother's male companion, Harry Blake, was a jolly fellow with a hearty laugh, and I liked him a great deal. Uncle Harry, as I called him, worked at Holts Wharf in Kowloon, and he and Aunty Olga, who worked at Credit Foncier d'Extrême-Orient on Hong Kong Island, often took the bus together from the ferry terminal back to the Taipo Road flat at the end of the workday. It wasn't unusual to see him at dinner.

Uncle Harry was a great tease, and I, unfamiliar with this form of affection, was a perfect target for his japes. He and Godmother came to check on me in the bath one evening.

"Look," he said, "you have a tail, and on the tail there are three feathers!"

I turned to look and saw nothing, and he explained that the tail appeared only briefly, and I must be quick to see it. They laughed uproariously at my efforts, but I didn't really understand that he was teasing. I believed in my tail for a long time afterwards.

Once, on an outing with Uncle Harry and my godmother, we stopped for tea (ice cream for me) at the Peninsula Hotel, where my mother had once run her gift shop. When the bill came, Uncle Harry said to the Chinese waiter:

"Give the bill to Robert, he will pay for it."

I was shocked; I couldn't believe it! The stern looks on the three faces staring at me convinced me that they were serious about this, and that was when I started to panic.

"But I don't have any money," I cried.

"Well, if you don't pay, you will have to go to prison," and that was when my eyes started to tear. The torment then immediately ceased, and all they could do was laugh at my fear and discomfort. It was a shattering experience, which I have not forgotten.

Uncle Harry loved his before-dinner drinks and would sometimes burst into song. In keeping with his personality, I remember joining him, singing the song "Who Killed Cock Robin". I learned the words well and sang it together with Uncle Harry with great gusto.

§

My blond hair was very fine and silky, and my godmother decided it needed to be "beefed up", which meant shaving my head bald and applying a special ointment to encourage robust regrowth. Mortified at

Godmother, Uncle Harry, and Babushka at 105 Taipo Road. It was here that I first met Uncle Harry and liked him because he was always so jovial. c1939

REMEMBRANCE OF FIRST HOLY COMMUNION

Robert Gatz

RECEIVED IN ROSARY CHURCH, KOWLOON, ON MONDAY, THE 10TH DAY OF APRIL, 1939.

PARISH PRIEST.

Remembrance of my First Holy Communion at Rosary Church, when Aunty Olga sponsored me and became my godmother. I don't know who my godfather was. c1939 the prospect, I protested, to no avail. Uncle Harry took me to a barber shop and waited for me while the wretched deed was inflicted.

Sobbing and feeling violated at the end of this ordeal, I hurriedly placed my toppee (also known as a pith sun helmet) back on my head as we left the barber shop, which I was loath to remove, except at bedtime. Insult followed injury, with the daily ordeal of having disgusting-smelling dog mange lotion rubbed into my scalp. I refused to leave the house for a few weeks afterwards, until my hair started to show some significant growth. At the end of it all, it made no difference, as the new growth was no different to what it was before the whole wretched ordeal started.

§

Most days, I hardly saw Babushka, as she was usually cloistered in her room with her pet terrier. Because of the language barrier, there was very little conversation between us. I filled my time on the wraparound verandah, watching people on the street down below and all the traffic whizzing by. I took to writing down all the license plate numbers in a notebook and recorded what seemed like millions of numbers by the end of a week. When that became boring, I tied a string to a Dinky toy car and dragged it around and around the verandah, mouthing engine noises.

Babushka had a silver fob watch; I think it originally belonged to her deceased husband. Even when fully wound, the watch would only

FOB Watch portraits from Godmother. Presently in my son's possession.

run when vigorously shaken, but then would stop after a few seconds. When Babushka gave it to me to play with, I too found shaking it did not fix the problem. I then had the idea of fastening the watch to my toy car, which I dragged behind me for several hours, going over the bumps of the tiled floor. This prolonged jarring of the watch worked a miracle. When I had enough of dragging the car around me, I stopped and found that the watch was ticking along by itself without stopping. I then lost interest in the watch, and Babushka, without me being aware of it, quietly retrieved it. Many years later, when Godmother died in Australia, this watch was bequeathed to me, and I have since given it to my son as a treasured heirloom.

With Mother in Matilda Hospital for the Summer

By this time, the void in my life was starting to get to me—I couldn't understand what was happening to my mother; nobody would tell me anything. "Where is she?" "What is she doing?" I don't know if my health was being compromised, but I do remember that I was beginning to feel listless and disinterested in my surroundings. Perhaps my asthma was acting up; perhaps it was loneliness and lack of stimulation that was the cause of my condition. I desperately needed my mother to restore balance in my life!

Imagine my pleasure and surprise when the next thing to happen was finding myself in Dr. Uttley's car on my way to Matilda Hospital. To me, that was a sure sign Mother was getting well, and before long, we would all be together again in our home on Nathan Road: Mother, Margaret, Julie, Ah Kai, Ah Ngan and Rexie. I don't remember what send-off my godmother gave me on my departure from her flat, but I don't think it bothered me too greatly.

Unfortunately for Dr. Uttley's car, I was prone to motion sickness. The winding Old Peak Road combined with the strong odour of leather didn't help the situation, and before we arrived at the Matilda Hospital, I had messed up the interior of the car. I don't know if I felt embarrassed or not, but I was certainly very glad to get out of the car as soon as possible.

By then, I hadn't seen Mother for over three months, but happily, we were soon going to be together. With my few belongings in a brown paper bag, I was taken to the reception desk, where it didn't take long to register me. I could barely see the top of the desk but couldn't wait to ask where my mother was; I had to see her, I told the nurse.

Mother's hospital bed was situated in an enclosed verandah near some large windows, which brought in a lot of light and fresh air. Through the large windows was a stunning view of the ocean in the distance, with Aberdeen and its harbour to one side and Pok Fu Lam on the other, with lush greenery in between. There she was, and I ran to her, crying out:

"Mummy, where have you been all this time? I was waiting for you every day. I missed you so much."

Mother must have tried to be as cheerful as she could, gazing at me with her beautiful blue eyes with me hovering restlessly by her bed from time to time; she couldn't even gather me in her arms to hug me, she lay so helplessly. I had barely recovered from the ordeal of the winding car ride, so I wasn't too demanding on Mother. She was lying quite still and wordless on her hospital bed, trying hard to give me a smile, when I burst out again:

"Mummy, I want you to get up; I want you to tell me you still love me."

I couldn't understand why she was unable to respond to any of my desperate appeals.

Then the duty nurse who was accompanying me lifted me up onto the elevated bed, so that I could be closer to her head. I think it was about then that I became confused and then horrified to realize that Mother couldn't move; I wept, and Mother must have struggled to cheer me up!

The diagnosis for Mother's illness was described as 'Influenza Transverse Myelitis of the Cord at level of 2nd and 3rd Cervical'. Only in exceptional cases is this affliction considered fatal, and this was to be one of those exceptional cases.

§

By the end of the first day, I was emotionally exhausted and still recovering from the recent carsickness. It was time for me to settle down and have supper, which the duty nurse took care of. I was assigned a bed at the end of a smaller ward, and after a bath and with clean pyjamas, I was happy to be tucked into bed by caring hands, with the mosquito netting carefully sealed around the bed.

It had been decided by the hospital authorities that I would be kept in the hospital for the duration of the summer for some rehabilitation and tonsillectomy in due course, under the care of Dr. Little. Also, it might have been considered therapeutic for my mother to be able to connect intimately with her very young son for as long as possible.

What a joy it was to be close by Mother after such a long absence, and to be the centre of so much care and attention from the nursing staff. The medical orderlies pampered me greatly as they gently nursed me back to normal health and strength on a day-by-day basis. I had full access to my mother's bedside, and it must have been exquisite pleasure for her to have me nearby. Unfortunately, or perhaps fortunately, I was totally unaware of her serious medical condition, which would take her life by the end of the summer.

At times, Mother asked me to massage her fingers as her hand rested on a rolled towel, and I would shriek with delight at the slightest sign of any movement, real or imagined. Oh Mother, how I weep for you now as I write my story.

The hospital nurses and orderlies were all so good to me during my extended stay in the hospital, and I well remember Fred, a male orderly, who used to amuse me by making exquisite Plasticine models of animals. The Plasticine came in multi-coloured round strips wrapped in cellophane and when new, would pull apart like cheese sticks. I treasured these figurines and happily showed them to Mother on my frequent visits to her.

The staff would also take me out onto the grounds when the weather was good. I walked around and even sat down to rest on a lawn chair and absorb the sunshine overlooking the magnificent view of Aberdeen and the glistening ocean waters beyond. I think we even walked around to the nearby nursing staff quarters, where I was entertained and spoiled from time to time by off-duty nurses in the nearby staff quarters.

One day, Mother said to me, "Robbie, make me a nice happy picture, with lots of colour. I know you can do it."

I got busy with wax crayons, working on a clean white sheet of paper. Happiness to me was a shining sun in a blue sky, and different coloured flowers in a field. I was rather clumsy with figures and animals. When I was finished, I turned to Mother and said:

"There, Mummy, what do you think of this?"

"It is beautiful," she said. "Bring it closer to me so that I can see it properly."

I climbed onto her bed and proudly explained the picture to her.

Courtesy Moddsey, gwulo.com

This was one of the passageways in Matilda Hospital from where one had a wonderful view through large windows of the ocean in the distance, and Aberdeen down below. Mother's bed had been wheeled out here so that she could enjoy brightness and see some of the view. This was where I found mother without realizing, initially, that she would be totally immobilized. c1930s.

gwulo.com

This was my first introduction to Matilda Hospital when Dr. Uttley drove me there. I was so looking forward to seeing mother again. c1930s.

"I like your sun in the blue sky, Robbie, my darling. You are so clever. Come give me a hug."

It was difficult to give her a hug, so I just snuggled as close as I could near her head and looked at her intently while admiring her beautiful blue eyes, almost for the first time, quietly wondering about her condition.

Then the nurse came and helped me off the bed. She also admired my artwork, and with that, I proudly paraded around the ward displaying my masterpiece to other patients.

I also remember vividly during the early stages of my stay I would wake up at night with frightful nightmares, and Fred, the male orderly, was often there for me, giving me assurance and comfort. Early one morning, visible through an open window across the room from my bed, the sky was a beautiful purple colour, tinted no doubt by the early blush of dawn. Suddenly, I saw an apparition within the window frame; it was the two-headed Jack of Spades grinning grotesquely at me as he gyrated continuously. I was transfixed and tried several times to erase it from my vision by blinking and rubbing my eyes, but without success; that was when I cried out with fear. Miraculously, Fred was by my bedside in a flash and pulling out the mosquito netting, he held me tightly in his arms. Sobbing and trembling with fear, I turned around in his arms, pointing toward the window, but the apparition had disappeared. He soothed me for a while until I had calmed down, and he then proceeded to tuck me back under the covers and tucked the mosquito netting thoroughly around the mattress.

During another night, I found myself so tangled up in the mosquito netting around the bed that I totally panicked. In the dark, I couldn't find which end of the bed was which. The duty nurse must have heard my distress, as she had to rescue me by extricating me from the tangled mosquito netting and bed covers, as well as calming me before tucking me back between the straightened sheets.

During this period, the medical staff decided that I needed to have my tonsils removed. This procedure was performed by Dr. Little, and the best part of it was afterwards being encouraged to sip ice cream to accelerate healing from the surgery.

During the last week of my stay in the hospital, I felt intensely happy, oblivious of the tragedy ahead of me. I am sure Mother wanted desperately to reach out to me and prepare me for the inevitable, but how could she explain such a serious topic to such a young person without causing me fear? Father's absence in my life, and his death, meant nothing to me, as I was too young to even be aware of the significance of that loss.

Did Mother talk to Margaret, my oldest sister, with any final words of advice and wisdom? As a young teenager, she too was greatly disadvantaged. I do not recall that my siblings ever visited me during my stay in the hospital that summer. The only thing Margaret mentioned to me, a good few years later, was that Mother was concerned that her children would end up being separated from each other. But separation was already a fact of life and would intensify with the passing years.

The day of my discharge from Matilda Hospital came just before the start of the school year, probably in late August or the beginning of September. Before leaving Mother for the last time, we shared a brief conversation:

"Mummy, when are you going to come home? I am going to miss you." Mother tried her best to assure me that all was going to be all right.

She said to me, "I want to see some more of your drawings when I come home. Promise you will make them for me? Then together, we will do a lot of painting, just you and I."

Looking at Mother, who seemed so helpless, it didn't seem very encouraging to me. Seeing me hesitate, she continued, "You also have to go to school, Robbie, my dear, and then when I come home, I promise, you can tell me all about it." For some reason, I felt a little better after these encouraging words, especially when she promised that we would be together again.

§

Without knowing what was lying ahead of me, I was probably excited with the coming changes and had no reason to feel insecure, naturally thinking, with the innocence of my age, that Mother would become well with time, and that once again, we would all be together again in our home. Alas, I was never to see her again!

Too young to have acquired any survival skills, my fate would soon be in the hands of total strangers, many of whom were well-meaning, but nevertheless remained as strangers during the crucial formative years ahead. Solitude in the next phase of my life would become a familiar companion. Due to circumstances beyond my control, at times, it even provided me a convenient escape from reality.

CHAPTER 5:
RESIDENCY IN DGS

Full-time Boarding School

1939 – 1941

It was probably my godmother who registered me as a boarder in DGS. I was given no explanation, nor did I seek any answers; I just knew that I had no choice, so I didn't dare question the decision. I consoled myself with knowing that Mother would be home soon anyway, and that my life would then revert to what it was before.

As I got off the bus at the junction of Jordan Road and Nathan Road with my godmother to walk the remaining distance to the school gates, I can remember looking up at the flat on Nathan Road, where not long ago I had spent many happy days with Mother. When the school gates closed behind me, I had no idea that the chapter of my entire previous home life was soon to be effectively closed.

At the Diocesan Girls' School office, I was introduced to Miss Gibbins, who had been appointed headmistress about a year previously. As I recall, she was a no-nonsense type of person, but she wore a ready smile. She wore spectacles, had some freckles and light brown, wavy hair, cropped short, which was tinged ginger. I was then turned over to the matron-in-charge; I think the matron was Eurasian and fluent in Cantonese. I recalled very little interaction with this lady, perhaps because she was more occupied with the girls under her charge and had little time for the few young boys.

I was not aware what arrangements had been made with the school for my welfare; I was also not aware what information about me had been provided to Miss Gibbins. I was too young to be entrusted with personal information about myself and my family; I was totally dependent on adults, all of whom were a confusion of strangers to me by this time.

After I said goodbye to my godmother, the matron led me to the boys' dormitory on the second floor. There were five or six iron-framed beds in this small dormitory, each with a heavy cotton bedspread over the mattress and a single pillow. I was assigned a bed and a place to stow my few belongings.

Being a boarder was going to be a totally new experience for me, but it did help that I was somewhat familiar with the school from my previous attendance as a day student. I was quite relieved to note that there were a few other boys about my own age also registered as boarders, and some of whom had older sibling sisters as boarders.

§

The following incident occurred about a month after England had declared war on Germany at the beginning of September 1939, and when, in Hong Kong, Germans and other enemy nationals were still being rounded up by the police for internment.

It was during recess one afternoon that I observed Ingeborg Warild being mercilessly bullied and tormented by fellow students in the school courtyard because they thought she was a German national. She was about two years older than me, but I thought she was being horribly treated. I remember her as a tall and lanky girl, reduced to tears so that teachers had to eventually intervene and put a stop to the bullying. When it was discovered that she was Norwegian, the apologies were profuse.

I don't know why, but this mistaken identity bothered me. I didn't think it had anything to do with her name, so then what was it? Maybe it was her appearance as blonde and blue-eyed? By appearance, Ingeborg might have been compared to some of the German families being rounded up. At that age I knew nothing about European racial differences, but only understood that England was at war with Germany and that English soldiers were killing the Germans.

Mother's Death

I was to become an orphan with no identifiable home address in the colony. Godmother had no legal responsibility for my welfare, and was probably reluctant to assume this responsibility in any event. I was totally unaware of my extreme vulnerability.

The fateful day when I was called to see the headmistress, Miss Gibbins, in her office, I had no idea what the meeting was about. I thought maybe Mother had come to take me home and went eagerly, full of optimism. I remember seeing Miss Gibbins standing alone by her desk, waiting for me. She sat me down and told me my mother had died in the hospital. "And you must attend her funeral, Robert. It's this afternoon, at Happy Valley Cemetery, you must get ready to go out." I was stunned and disbelieving!

My distress was total, and my response to Miss Gibbins showed it. "Mummies don't die! My mummy is only sick, and she will get better; she told me so! She promised," I cried!

The headmistress stood silent and motionless next to her desk. Perhaps she was preoccupied thinking through the long-term ramifications of the news. Whatever the reason, she offered me no comfort.

I was beside myself with anxiety, totally confused and not knowing whom to turn to. There was nobody. My sisters were so foreign to me that they never even entered my mind in this time of great distress. My godmother also could not have been farther from my mind. The only one who could have consoled me was Ah Kai, but she too had dropped out of my life.

Newspaper Announcements of Mother's Death:

South China Morning Post September 29, 1939, Page 10.
Announcements – Deaths:
TATZ-ROJDESTVIN. Mrs. Antonina Tatz-Rojdestvin, in Matilda Hospital, at 7 a.m. on September 28, 1939.

SCMP October 3, 1939, Page 10. Announcements – Acknowledgements
The Committee of the Russian Orthodox Church Community at Hong Kong and the closest friends of the late Mrs. A. Tatz-Rojdestvin wish to thank the Trustees, the Medical Superintendent and nursing staff of the Matilda Hospital for their devotion and care – and all friends for floral tributes and attendance at the funeral.

Mother died at age 35, after ailing for about six months. Her death certificate listed the cause of death as "influenza transverse myelitis of cord at level of 2nd and 3rd cervical." All of Mother's hard work to support herself and her children very quickly came to nought!

What happened afterwards is a blur in my memory. Someone—I don't recall who—escorted me from Miss Gibbins' office at the school to the cemetery in Happy Valley; nor do I recall the bus trips, the ferry crossings to Hong Kong Island and back or the tram rides. The excitement of the ferry rides I used to enjoy with Mother was not there; I looked at the harbour and at all the usual activities, and yet I saw nothing.

All I remember is descending from the tram outside the Jockey Club building at Happy Valley and being led across the tracks through the cemetery gates. We turned left and, before long, I found myself among a small gathering in front of a polished wooden casket. When I reached the casket, I was told Mother was inside.

During the ceremony, I was encouraged to be the first to throw some earth onto Mother's casket as it was lowered into the ground. I felt numb, and oh so lonely!

"What now?" I thought. What now, indeed! I was lost in my thoughts all the way back to DGS. As we approached the school, I was woken back to reality by the sight of the gates. My first thought was whether I would be in time for high tea. This was my peculiar form of denial under these stressful circumstances.

The first night was dreadful. The nightmare I was grappling with was that I would never see Mother again—not now, not ever! Totally unimaginable! I was haunted by the memory of chanting and incense smoke, which I could still hear and see as I was lying on my bed in the darkness; of the throwing of flowers onto Mother's casket; of dropping the first clump of dirt; thinking of Mother's closeness, recalling her helplessness as she lay on her hospital bed. Then came thoughts of the happy times brightened by the sun and colourful flowers. How could she abandon me when I still needed her? As heavy rain is the norm after the passage of a typhoon, my tears flowed copiously until, utterly exhausted, sleep brought relief.

Adjusting to Life as an Orphan

Accompanying my dreams for a long while later, I had strong memories of Mother. She was petite, brunette and had beautiful blue eyes. Being an artist, she loved her colours. In the summer, she used to wear a light white dress with splashes of red. She looked like a princess when she wore her beautiful black evening dress. She also had a cream-coloured dress, which can be seen in a photograph taken on the roof of our home. The one I liked best was her two-piece military-style linen costume. In cooler weather, she wore a plaid jacket over a purple sweater, which is the image shown on her naturalization certificate and

British passport. This is the final image of my mother, as I remember her.

It was going to be a daunting challenge to adjust to this new world on my own without the safety net of a parent or a home. I found myself totally dependent for my welfare on the adults in authority in the DGS. Who else was there?

My teachers, the matron in charge of boarders and the headmistress of the school presumably did their best for me, and with so many children in their care, they could hardly give me any special attention, as far as I can remember. They were kind where they could be, but not warm or affectionate. Left largely to provide for my own comfort, this was when I started the habit of sucking my thumb.

I was obliged to wear a black armband every day for about a month. The whole school could see that I was in mourning, and word quickly spread that I was now an orphan, which attracted some unwelcome attention from my classmates. I hadn't come to terms with my loss and didn't know how to respond to curious and personal questions from pupils whom I was normally in contact with during the day. In the early days, this period of adjustment was stressful, and I was often exhausted by bedtime.

My eighth birthday fell on a Wednesday that year, not long after the cemetery experience. It was a normal school day, and I was probably unaware of the occasion, as it came and went without any acknowledgement or celebration, unlike last year. I was not even aware of my ninth and tenth birthdays while living at this school. I felt no resentment, as there was no one to remind me about these events, and that was the way it would mostly be until my 21st birthday, 13 years later.

The life of an orphan boarder at school had a completely different tenor to that of a day student when Mother was alive. I had to learn quickly to suppress my emotions. Coping with grief during the day was usually not a problem because my mind was on classroom activities. At noon, we met at the refectory for a luncheon snack before returning to classes, and sometimes there was time to run around the school's sports field for a few minutes before the school bell rang.

Miss Gibbins, the headmistress, lived in private residential quarters on the school property and took her meals at her own table in the boarders' dining room. She usually said grace before we were allowed to sit down and eat. Our meals often included sausages, mashed potatoes, mutton, boiled spinach, porridge, egg custard, stubby little sausages and rice. Drinking water was filtered inside a cylindrical porcelain tank, which seemed to take ages to drip through the filter elements.

I can only recollect the name of one other boy of similar age in my dormitory — Leslie Morley, who seemed pleasant enough. The evening routine was set: high tea, homework for older students and a shower before bed. The shower facilities were not segregated, and the youngest students, the boys, showered first. The older students would follow after they had done their homework. After showering was weekly grooming at the hands of the matron, which I dreaded; the matron trimmed fingernails and toenails very short and invariably drew blood.

I enjoyed the daily gymnastics on the school playing field before start of classes in the morning (weather permitting). At other times, I watched senior girls playing grass hockey. However, I don't remember much about participating in any sporting activity myself, not even at times of inter-school competitions. This might be attributed to lack of sponsorship. There might have been games for the early grades just to keep them amused while the seniors competed among themselves mightily.

After classes on Fridays, I watched from a distance as excited boarders prepared to leave school for the weekend with their parents. I watched again on Sunday evenings when the boarders returned and envied the boys their parents' indulgence, hugs, kisses and affectionate chatter. I well understood why some of the younger boys were in tears as they said goodbye for the week, hating to tear themselves away from their parents, even for only five days.

§

At the beginning, I often felt lost when everyone went home on weekends, but by the time summer holidays came around, I had adjusted to the solitude and learned how to amuse myself. I spent long stretches of time alone, but these were balanced by happy events and certain periods of company and pleasure. I soon got used to these swings in my life and became skilled at masking disappointment. I developed a certain stoic acceptance of my changed circumstances. There were ways I changed that I did not even recognize at the time.

If my life was directionless and uncertain, it was at least safe and not uncomfortable; I had food to eat, clothes to wear, a bed to sleep in and a roof over my head. My horizons were admittedly narrow, and my days a blend of dull routine among other people and bored solitude, but that would soon change. I did not know this, of course, so I was not anxious within myself or concerned about the unknown future.

Outside (or non-school) Activities

Not long after I was registered as a boarder, I became a member of the boys' choir at St. Andrew's Anglican Church on Nathan Road. I don't know how long this lasted, but it gave me something to look forward to on Sundays and the company of other boys around my age. I loved wearing the red cassock and white top and enjoyed the singing tremendously. I always sang the hymn "All Things Bright and Beautiful" with great gusto. The local dean and bishop, Dean Rose and Bishop Hall, were frequent visitors and were aware of my situation from Miss Gibbins.

§

I was in the First Kowloon Cubs, which met every week during the winter months in St. Andrew's church hall on Nathan Road, across from Whitfield Barracks. Akela, our leader, was a soldier from

This is St. Andrew's Church post WWII. It hasn't changed much from the time that I used to frequent it and the church hall for various activities. The highlights of my memory were the 1st. Kowloon Cubs weekly meetings and singing in the choir of this Anglican Church. They were enjoyable interludes in a desperately lonely period of my life. c1940.

Whitfield Barracks. I was in awe following in the shadows of a real soldier, and felt great wearing my own uniform, a long-sleeved grey shirt, khaki shorts, knee-length stockings and a red neckerchief held together at the front with a loop of leather called a woggle, all topped with a green beanie cap. I never found out who contributed or donated this gear to me.

Our weekly meetings were usually held in the evenings and started and ended with the Grand Howl. Our leaders were all named after characters taken from Rudyard Kipling's ***The Jungle Book***. We had Akela, and sometimes other leaders called Bagheera and Baloo joined us.

As a Wolf Cub, I learned the Cub Promise, Motto and Law, and above all, to be chivalrous to women in all situations. I also learned how to fold clothing (a skill I unswervingly practice to this day), how to tie different knots, the significant elements of the Union Jack flag, how to polish my shoes, knowledge of courtesy and code of conduct. I particularly enjoyed playing "Capture the Flag" in a darkened hall at the evening meetings. Spiritual fellowship was achieved by my participation in St. Andrew's Church services. I knew the national anthem and readily sang it whenever required.

It was here that I was introduced to the origin of the Union Jack flag, which was a combination of the following: Cross of St. Andrew with a white saltire on a blue background; Cross of St. Patrick with a red saltire on a white background; Cross of St. George with a red cross on a white background. All this cemented my allegiance to Great Britain.

The Cubs had no regular meetings in the summer months, but the troop did have at least a one-day camp at a selected beach, either off the Caste Peak Road or at Kowloon Bay. On one beach outing, an undercurrent had dragged me out of my depth, and I was just starting to panic when a strong arm from behind suddenly scooped me back to safety; it was Akela, my hero!

§

Elizabeth Rodgers was a friend of my mother, a kindly lady who may have been Russian or had Russian ancestry. She and her husband lived in a house on the Peak, I think on #11 Coombe Road. Two of their sons were in college in the USA at the time.

Robert Augustus Rodgers was an American stockbroker or banker, and they used to entertain lavishly with weekly Sunday brunches. One Sunday, I found myself at their house on the Peak, where Elizabeth Rodgers gave me a lot of attention. I happily mingled with her guests, and made the acquaintance of Christine Corra, who was surrounded by a small group of young naval officers; they had all just returned from an early morning horse ride in Pok Fu Lam.

At some point, Mrs. Rodgers asked me to go to the chicken coop at the back of the garden to gather some fresh eggs, but I was terrified of a ferocious gander which was blocking my path. Every attempted move I made resulted in a charge by the gander. Eventually, Elizabeth Rodgers came out, and to my amazement easily subdued the gander by holding its neck. The next problem I encountered was the reluctance of the hen to surrender her eggs in the nest, but again Mrs. Rodgers showed me how to solve this problem. She made me feel very special, knowing my circumstances, and because of this it was natural for me to feel strongly drawn towards her. A few happy hours which I relished intensely before returning to the boarding school.

These, and many other similar happy outings, inevitably ended in heartbreak for me. The return to emptiness at DGS at the end of each outing was always a forlorn experience, an anti-climax to a very happy day. In this case, I wouldn't doubt that Elizabeth Rodgers probably sent me back to DGS with a care package, not knowing the futility of this gesture. Being repelled by the silence of an empty school on weekends must have been 'deafening' after a fun-filled day.

§

Outside the school gate and across Jordan Road was the Kowloon Bowling Green Club. During the summer holidays, I often spent hours watching the adults at their games when I had nothing else to do. Behind the club property and across the road was a block of terrace homes, occupied by expatriates, mostly from England. I played with some of the children living there and remember three in particular: Keith Armstrong and Dorothy and Barbara Keates.

I had enormous fun playing with this group of children, and I always looked forward to crossing Jordan Road to join them. We played hide-and-seek, tag, marbles, skipping and hopscotch. We used various means to pick who was 'it' or to form up teams: eeny meenie miney moe, rock paper scissors, one potato two potatoes... and when all was settled, we would get into our game.

Some of the older boys used to take great pleasure teasing the girls; on one occasion, they caught Barbara and tied her with rope to a downspout, with the rest of us cheering. When the call came from parents to come home for tea, Barbara was abandoned, still trussed up and totally helpless. Dorothy freed her sister and they went home together, with Barbara none the worse for her experience.

This is one incident that I have never forgotten: Keith had invited some of us into his house to play with his toys. He had a staggering number of them: Dinky toy cars, combat soldiers of various kinds in lead and rubber, model fortresses, a HO model railway, toy weapons and games galore. When teatime came and we all got ready to go home, I had the idea to take one of Keith's soldiers with me so that I could have somebody to talk to when I went to bed that night. I was exiting through the front door when one of Keith's friends (a bossy older boy) jumped me and snatched the soldier out of my hand, accusing me of stealing. I tried to protest, but Keith only looked at me in bewilderment. He said nothing as I tearfully hurried out of his house, full of shame and mortification. A very unpleasant confrontation in my life—I felt so wretched over this incident that I never returned to play with that group of children again; I sought solace in solitude.

§

The Dutch consul for Canton and Hong Kong was Mr. Weirink. He and his wife were the godparents of my sister Julie, who was born in Canton. It is possible that they first became acquainted with my parents when the family resided in Shameen, before they came to Hong Kong. They had a residence in Hong Kong at the Repulse Bay Hotel.

It was customary for the Dutch consulate to annually host a Christmas party for the small Dutch business community and their families. I attended one of these events in December of 1939 or 1940. The hall was full of children of all ages and their parents. I was awestruck by the magnificence of the large, decorated Christmas tree. It is odd that the display of food never registered in my memory, though there must have been stacks of food being served, that would have been a welcome break from the monotonous food of the DGS.

After much excitement and anticipation, St. Nicolas arrived, accompanied by Black Peter, who had a blackened face, wore ragged clothing and carried a sack of coal lumps for the naughty children. To everyone's relief and delight, no naughty children were identified, and everyone happily received a gift from Santa's hands. Black Peter helped to pick up all the inevitable debris after the gifts had been opened. I wish I could remember what gift I received.

§

In the summer of 1940 (or late 1939), I believe that Bishop Hall, Reverend Dean Rose and Miss Gibbins made efforts to offer me for adoption. I was eight years old, but slight in stature, so I looked six. Two parties expressed an interest in my adoption: an American couple visiting Hong Kong, and two sisters, members of a prominent local philanthropic Eurasian family. This information was gleaned from my godmother many years later.

The interview for adoption by the American couple took place on Hong Kong Island at the Church Guest House, or at Bishop House, I am not sure which one. I have no memory of who the people were or what they looked like.

The other interview was during a weekend spent at a house in Kadoorie Avenue, which was somewhat different. I can still remember being interviewed by two Eurasian ladies. In this house, mostly Cantonese was spoken, and there appeared to be quite a few Chinese servants in evidence. For some reason, I was reluctant to engage with them in the Cantonese language. I can remember the servants helping me to have a bath the first evening; I guess to make me more presentable. Apart from that, I have no further memories.

According to my godmother, who related the above activities to me, as part of the process for adoption, consent for adoption was sought from the oldest and closest available family member; in this case, it was Margaret. Apparently, at 14 years old, she declined to give her agreement, citing that Mother's last wishes to her was that the family should not be separated. So, both adoption efforts came to naught.

§

More than anybody else, the Kriloffs gave me the most pleasure in my life, up to the time they were evacuated to Australia in the summer of 1940. It helped that they lived very close to the school, even closer than I did when I was living with Mother on Nathan Road. This is reinforced by the memory of my seventh birthday celebrations, together with Mother, in their home.

The Kriloffs, Costia (Constantine) and Lydia, were a childless couple and very good friends of my mother; we often visited their flat, located above the Evergreen Store at the junction of Nathan and Jordan Roads.

Costia used to roar with laughter when I pretended to arrest him and anybody else in view, using his unloaded service revolver with the command "Stick 'em up, and follow me behind!" Of course it never occurred to me how ludicrous it was to expect my prisoners to comply with my command under such circumstances. To humour me, and meanwhile giggling, my 'prisoners' would obediently follow along in single file with their arms raised above their heads. Costia thought it was hilarious, and often related this story to his colleagues in the police department. I was eight years old at the time.

In the early months of my bereavement, they often took me to their flat to spend some of the weekends with them; those were precious times. On occasion, Uncle Costia or Aunty Lydia would turn up at the school and, to my great delight, walk with me to their nearby flat. I would dance around them ecstatically, practically skipping along the sidewalk ahead of them all the way to their flat and racing up the stairs to the front door. I knew I was going to be pampered, at least for a short while!

I used to play endlessly with two objects in the Kriloffs' flat when I overnighted with them: a series of detachable carved miniature ivory elephants, intertwined trunk to tail and mounted on an ebony stand, and a silver cigarette case in the shape of a flat-bottomed barge. It had a detachable lid over the cargo hold for storage of cigarettes. I had hours of fun with these two items.

Costia Kriloff also possessed an air rifle that he let me use to shoot at targets on his verandah under his supervision. The bullets were lead pellets, at one end of which was a tiny coloured feather. The targets I used were some of my lead soldiers, which I used to line up against the verandah wall. Resting the rifle on the back of a chair, I would fire at these targets, which just about dislocated my shoulder each time I squeezed the trigger. Most of the time, Costia had to load the rifle for me.

My joy in spending time with them was somewhat allayed at the end of each visit by the letdown of being returned to the school and left to my own devices again, but the memory of the enjoyable time we had together made being alone a bit more bearable.

Under government orders, Lydia and Vitaly Veriga's first wife were part of the mass evacuation of British women and children to Australia in the summer of 1940. After they left, I hardly saw Costia, and I believe he was very busily involved with police work and tracking down illegal activities. He left Hong Kong by boat for Singapore just before Japan invaded Hong Kong. Arriving in Singapore, his ship was bombed and sunk, but he made his way to Australia on another ship. Subsequently, he was recruited by the British Army and spent most of his wartime activity in Burma. He and Lydia returned to Hong Kong after the war before their final retirement in Australia.

I understand Margaret, who was about 15 years old, also visited the Kriloffs from time to time, but I cannot recall ever seeing her come to visit me at the nearby boarding school where I was living. Due to our previous lifestyles, it was unlikely that I missed my sister, as to all intents and purposes, she was almost a stranger to me.

According to Margaret, Costia took custody of some of our family assets after Mother's death; before he left Hong Kong, he had assured her that the assets were in safekeeping at the Hongkong and Shanghai Bank (HSBC), where Mother used to have her financial account and security box. Whatever happened to these assets is unknown, and as far as I know, Margaret never pursued any investigations.

§

Mr. and Mrs. Tipple were friends to Aunty Olga and Harry Blake, and that was how I was introduced to them in their home at Red Roofs near Kai Tak Airport. It was sometime in early 1940 that Godmother took me for a visit to the Tipples at their home in Red Roofs. I was drawn to Mrs. Tipple from the start. She was a no-nonsense type, but very kind. She must have known about my circumstances, but she never intruded into the private space that I had built around me. When I was in their company, I felt safe to lower my guard and play and laugh whole-heartedly with Lesley and Berry, her two children about my age.

They had a cook boy (Ah Sun) and four other servants, including Fa Wong (a gardener). There was a large garden where they grew all sorts of vegetables and reared geese, goats and chickens. Sitting in an enclosed porch, a stunning view of Lion Rock was visible in the distance. It was all so magical!

Lesley and Berry were good fun to be with. I fondly remember the good times I enjoyed playing together with them. Sometimes we ventured into the concrete pool that at one time was a fish pond. Brutus was their large Alsatian dog, and I remember how terrified I was of him.

Lesley and Berry were both attending DGS as day students at that time, but I do not recall seeing them there; they may have been in a different grade. Their father, who was the chief engineer for Kowloon

Motor Bus (KMB) at the time, always arranged for the bus driver to drop them off right at the DGS school gate.

In the evening, at the end of a marvellous visit to Red Roofs, my godmother would drop me off at the gate of the boarding school, and then she would continue on the way to her home at Taipo Road. I would have had my high tea earlier and was ready to be sent off to bed upon my arrival. The feeling of loneliness lingered for a while, but at least I was safe, and I relished the memory of each happy hour I had been enjoying during the day as I waited for sleep to come.

In the summer of 1940, Dorothy and her two children registered with the government for evacuation from Hong Kong to Australia. They left Hong Kong on the Empress of Japan with a stop in Manila for two weeks, and then two months in Baguio, before finally arriving in Sydney.

This emergency order for evacuation of women and children from Hong Kong to Australia in June 1940 had a 36-hour notice clause. Once you received your notice, you had to respond almost immediately.

If Mother had been alive, she and her children would have been required to comply with this order, as all four family members were British subjects. Margaret was 14 years old; Julie was just over 10; I was almost 9.

It was ironic that Mother obtained British naturalization, including for two of her children (I was a natural born British subject) in January 1939, which entitled all of us to protection by the Crown, and then to have Mother die the same year in September, leaving her three children orphaned. Without sponsorship, there was no recourse for dependent children to comply with the evacuation order. Because of this, our orphaned family endured the subsequent battle in Hong Kong and Japanese occupation over the next three years and eight months. This omission was tacitly acknowledged by the Crown at the end of the war when all three of us were repatriated to the UK to benefit from a year's rehabilitation; the same as was accorded to all Released Allied Prisoners of War (RAPWI).

Other Activities (within the school grounds)

To help fill voids of emptiness during summer holidays, I used to wander over to see what the school's Chinese groundskeeper was doing in his vegetable patch at the back of the school grounds. It was during one of these wanderings that the gardener helped me create a crude little celery patch. Under his guidance, I carefully tended my celery but was dismayed that it so easily attracted aphids, and his was free of this pest. It was during one of these times that Yolanda walked across the field to see what I was doing.

I know nothing about Yolanda except that she was boarding at the school for a short period that summer. She was about 14 years old and had a younger sister who I didn't know at the time was afflicted with Down syndrome. But in the short time she was boarding in the school, I found out that Yolanda had a vivid imagination, and was great at telling spooky stories. I was captivated by her tales as we walked together around the playing field, with me as her sole audience. I was sad when this short interlude came to an end.

During times when the school was out, the days were often long and tedious. I didn't dare wander far away, but I would sometimes sit on the curb by myself just outside the school gates, watching the flow of traffic. Disembarking passengers at the nearby bus stop always caught my attention, as I would longingly hope to see somebody I knew, somebody who might be coming to see me or take me out for a treat!

§

However, when school was in session and at times of inclement weather, there were interesting activities that I enjoyed participating in. We played many board games, and card games were always a favourite pastime; Snap was very popular, which was usually played between two children and full of tension resulting in frequent false shouts of unwarranted 'snap'. I built castles out of cards when there was nobody around. I enjoyed the challenge of using a steady hand to avoid collapses. Go Fish was another favourite, but which required multiple participants. Sometimes we worked with jigsaw puzzles; it was a good game to learn to work in a team. I spent many hours in these activities, sometimes interrupted by the welcomed offer of a glass of milk and biscuits. Of course, board games also included Ludo and Snakes & Ladders.

§

At school, when the weather was good, we spent some of our endless youthful energy playing games outdoors. Games like tag, hide-and-seek, and what time is it, Mr. Wolf? all required no equipment. Blind man's bluff needed only a blindfold for whoever was 'it', who had to guess the identity of players they tagged.

Competitive running races, often for a prize, were popular. A bit of rope or even a school tie could be pressed into service to get a three-legged race going. Sack races took a bit more forethought, but sprints could happen anywhere, anytime.

Games and activities involving a ball of any kind were especially popular. A person could play alone, with another person, or with a whole team. We played a great variety of ball games and sports, especially with adult involvement. Even alone, I would bounce a tennis ball off a wall for hours.

§

I had the whole school, building and grounds mostly to myself for 48 hours on weekends. So, I invented games and activities to entertain myself. I made paper airplanes and made them perform aerobatic feats. Sometimes I did the same with simple balsa wood airplanes made from a kit; these were powered by a single propeller driven by winding elastic band, a great technological leap forward.

I folded paper boats and floated them in puddles or in a flowing gutter after a rainfall, testing how much cargo the paper boats could carry before capsizing. I had hours of amusement in climbing trees and shooting targets with catapults from elastic bands, too.

With other children, I played hopscotch on a sidewalk or cement floor, hit and chased hockey balls on the grass field, played pickup mini-bags, also known as chucks or grab piece, and tin can football. I collected live insects in matchboxes, which got thrown away after the insects died and started to smell. Captured cricket fights were another messy pastime.

§

I collected few toys, as there were no personal storage facilities for hobbies in the boarding school, but I thoroughly enjoyed playing marbles, alone or against other boys. We drew the playing circle on whatever surface was on offer, sand or pavement. Each player would deposit an agreed number of marbles inside the circle and then from a distance of five or six feet, would toss a prized 'aimer' as close to the circle as possible. An aimer that landed within the circle remained in the circle, and the player tossed another prized aimer. With aimers positioned, players then took turns, trying to shoot marbles out of the circle, which then became a prize to keep. Naturally, aimers within a circle were prime targets for other players. It was vitally important not to carelessly pop your own prized aimer inside the circle!

§

I had the use of a pair of roller skates at one time; these were probably borrowed. I spent endless hours rolling on the sidewalks outside the school walls. My shoes didn't have the right sort of soles for the skates to firmly grip on, which caused regular involuntary and painful dismounts. But this was hardly a deterrent from so diverting an activity.

My sandals were often reduced to a dilapidated condition, which caused me certain grief. It was difficult to avoid tripping when the sole parted from the leather top; wrapping with twine around the toe didn't last long. I must have looked a sight at times, with my sandals or shoes flapping as I walked. Running under these conditions invariably ended in falls, resulting in scraped knees, cut hands and bruises galore, but nothing more serious. My abrasions were treated with Mercurochrome or iodine (I much preferred the former), and I carried on until new footwear miraculously appeared, seemingly out of nowhere.

§

As this period came to an end in my life, there was one more excitement which has always remained in my memory. It was the advent of the deployment of Canadian soldiers in Hong Kong.

On October 27, 1941, a battalion of the Royal Rifles of Canada and the Winnipeg Grenadiers had embarked on the troopship *Awatea* (a converted cruise ship) in Vancouver, and accompanied by HMCS *Prince Robert,* arrived 20 days later in Hong Kong on November 16. Coming off the troop ship at the Kowloon Wharves, the soldiers marched along Nathan Road to their barracks in Shum Shui Po. They were known as C Force. This caused great excitement for the population and attracted crowds to watch the parade. The added strength of the defence force gave Hong Kong citizens a measure of comfort and security without knowing how short-lived this would eventually become.

This was a memorable day for Hong Kong and its citizens. Imagine the excitement felt by the few boys in the classroom at the school, including me! Being in class prevented us from going out to watch the parade and to witness this historical event happening right here in Kowloon. But the next day, some of the boys were able to share the spectacle that they had witnessed with their parents.

By this time in Hong Kong, the population was frequently hearing the practice wail of air raid sirens, which was quite audible from our school. The talk in school the next day was all about the Canadian soldiers that parents had been discussing, the first-hand news that I was denied. Of course, by the end of classes, the parade was over—activities on Nathan Road had quietened down and returned to normal.

By this time, I had endured considerable solitude for about two years as an orphan. Because of this, at 10 years old, I had learned how to shrug off disappointments, and about the futility of harbouring resentments, at least in the long term. These traits and complacency have certain negative impacts as is it not the "squeaky wheel that gets the oil"? Without fear, anxieties or any expectations, I was about to leave forever my past life to imminently enter a perilous phase. In the meantime, my attention was totally absorbed by the presence of increasing military activities, which is pure adrenaline to any red-blooded boy. This was fueled by frequent air raid sirens backed up by the presence of uniformed Air Raid Wardens.

Wikimedia

Arrival of Canadian forces in Kowloon on November 16, 1941; the Winnipeg Grenadiers and Royal Rifles of Canada. Although I have no recollection, I might have been able to see some of the march pass on Nathan Road which was not far from the school.

CHAPTER 6: BATTLE FOR HONG KONG

Who is this Orphan Boy?

December 8 – 25, 1941

Tribute: I dedicate this chapter to the memory of William Sewell, without whose support these memoirs might have never been written. I am grateful for the shelter the Sewell family gave me in a time of extreme peril. I would also like to pay tribute to the Kennedy-Skipton family, for their generosity in accepting me, a stranger, in their home on the Peak, together with the Sewell family and other refugees.
Hundreds of books have been written about this 18-day period in Hong Kong's history, the following is an account of my experience as a 10-year-old.

For some time, Hong Kong had been actively preparing for the event of war with Japan. The Japanese army had captured Canton and had been positioned on Hong Kong's border in the New Territories since 1937. Air raid shelters and defensive pill boxes had been hastily constructed. There were frequent air raid drills and army manoeuvres. The navy too was active, especially the Motor Torpedo Boat Flotilla. The volunteer reserve forces were being bolstered and in constant training, together with the garrison militia. However, despite all these preparations, the odds were still far against Hong Kong.

The Battle for Hong Kong, also known as the Defence of Hong Kong, was one of the first battles in the Pacific, in the Second World War. Early morning of Sunday, December 7, 1941, a high alert was issued to all units of the defence force, and early the next morning, forces of the Empire of Japan invaded the British Crown colony of

Hong Kong. Simultaneously, Pearl Harbor was also attacked by the Imperial Japanese Navy.

The Japanese attack was met with stiff resistance from the Hong Kong garrison, composed of Hong Kong Volunteer Defence Corp (HKVDC), as well as regular British, Canadian and Indian military units. However, in less than a week, Hong Kong defenders had abandoned the mainland, and less than two weeks later, with their position on the island untenable, the colony surrendered. The following is an account of my experience during this short, turbulent period in the history of Hong Kong.

Day by Day Account of the Battle

December 8, 1941

This Monday was going to be a long day. At 8:00 a.m., I had just finished breakfast at DGS and was heading towards my classroom with the other boarders when we heard the baleful wail of an air raid siren. The sound was not new to us, as these signals had been a common occurrence for the past month or longer. As usual, we ignored it, putting it down to another routine practice. The school gates were open, and day students were arriving amid the usual Monday morning chatter. Abruptly, several fully uniformed Air Raid Precaution (ARP) wardens came through the gates and quite excitedly were warning everybody that this was not a practice alarm, but the real thing. Almost simultaneously, explosions sounded from the direction of Kai Tak Airport. The wardens then urged those students in the classrooms to shelter underneath the school desks and wait for the 'all-clear' siren; they then went off to talk to Miss Gibbins and the teaching staff about the developing emergency and the advisability of sending all the children home.

I watched as singly or in small groups, some sobbing in fear and some stunned into silence, children left the school in the company of family members or servants. Standing in the school yard, I watched Miss Gibbins and some of the staff hurriedly secure the school to protect it from looters, and then Miss Gibbins started loading her car with her personal belongings.

By then, the school had been evacuated; Miss Gibbins told me to go outside the gate and wait on the sidewalk. She drove her car out, stopped just past the gate, got out, locked the school gate behind her and got back in her car. She then drove away without another word, leaving me standing alone on the sidewalk, waiting for the next instruction.

I watched in bewilderment as her car sped away. She hadn't said how long I should wait, and this section of Jordan Road was awfully quiet at that time of the morning. I just stood there, next to a locked school gate, wondering who I was supposed to be waiting for. This moment is forever etched in my mind.

Unbeknownst to me at the time, when Miss Gibbins left me, she was hurrying to her designated first aid post, and eventually ended up at the La Salle Auxiliary Hospital on Boundary Street, not far from Kai Tak. She had registered with the Auxiliary Nursing Service (ANS) on September 20, 1941, and this hospital was to be her emergency post. Civilian casualties were starting to stream into the hospital because of the bombing.

Minutes ticked by and no one showed up and I began to think that probably no one was going to show up. I needed to do something, though I was anxious about leaving the only place that was home to me. If I walked away from the school, what could I tell people about me? By now, I had been an orphan for a long time. The only aunties and uncles that I knew were not family members and, anyway, I wouldn't know how to connect with any one of them. I didn't even know that the Kriloffs had already left the colony. Without relatives or even close mentorship living in Hong Kong, the only person who could be considered a relative was my godmother, Olga Robinson. Where was she? Who else could I turn to for help? How much longer had I to wait?

My godmother was living on Taipo Road and might have decided to wait out the developing emergency at home. Harry Blake had rooms in Tsim Sha Tsui, but I did not know where he would be at that time. I was located approximately midway between the two residences of people who might have been able to help me. I didn't have the means or ability to mount a search for either one of them, as this would entail a bus ride.

My own sisters were boarding in a convent on Hong Kong Island, whose location I could not have found on my own. Since I had little or no contact with either of them after Mother died, it never even occurred to me to think of them as a source of help; but anyway, they were not anywhere near DGS. I was more preoccupied with my own dilemma than to think about them.

All personal information about me was on file at DGS. I had no identification on my person. All I had on me was what I was wearing. If asked, all I could say was, "I live at DGS, where I am a boarder; I am 10 years old; I am an orphan; I don't know where to go or what to do!" At any other time, I could have given Miss Gibbins' name as someone who knew all about me. But it was Miss Gibbins who had just driven off in her car, leaving me alone with instructions to wait.

This inability to provide documentation to validate personal information about myself would haunt me throughout the war and into early adolescence. The embarrassment of being unable to fully introduce myself to interested strangers and provide information about my background caused me to withdraw further into myself. Was that why so many adults used to look at me with disbelief, wondering, "Can this boy be such a lost case?"

During my research for these memoirs, I came across a very interesting occurrence which had taken place about the same time in the Diocesan Boys' School (DBS), originally known as DBSO (Diocesan Boys' School Orphanage [sister school to DGS]). This occurrence was reported in the first postwar DBS Annual Report:

> *"Under fire by Japanese soldiers, Mrs. Hassard (she was 55 or 56 years old at the time), matron for the orphan boarders, shepherded a group of young boys (I believe there were about 11 boys) from the DBS, crossing the Harbour (in one of the few boats still available) on December 11. Mrs. Hassard's actions were commendable and very courageous, however, she would probably answer any comment of praise with 'I was just trying to get these kids to safety. Anyone would do that.' In this action, she was helped by one of the older boys, named P.A. Walker."*

I have absolutely no doubt that the boys under Mrs. Hassard's care felt protected. Where was the matron of my school?

In the distance, I could still hear what appeared to be sounds of explosions, probably Kai Tak still under bombardment, and this spurred me to action. Jordan Road was remarkably free of traffic and pedestrians at this time of the day, and I saw nobody I could recognize. Two blocks up, crowds of pedestrians were surging south on Nathan Road, toward Tsim Sha Tsui. I debated what to do and where to go, and eventually, since there was no one to ask permission to leave, I went the two blocks towards Nathan Road to see what was going on.

What I saw out on the street explained nothing to me. Hundreds of people streaming down Nathan Road, from the general direction of Sham Shui Po toward Tsim Sha Tsui. The crowd's fear was palpable, and all the faces showed distress. Where were they all going, and why? Some buses were still running, but pedestrians soon impeded vehicular traffic.

My Trek to the Star Ferry Terminal

Not understanding the gravity of the situation, I was not afraid. But I needed to do something, to go somewhere; the notion of waiting around all day and having nowhere to stay when night came did not appeal to me. I was, in fact, delighted that there would be no classes for the rest of the day (let tomorrow take care of itself). So, with no other options and still alone, I joined the growing melee heading south toward Tsim Sha Tsui.

At first, my greatest concern was that I would be in trouble with the school for wandering so far away by myself, without permission and contrary to instructions to wait, but I forgot about this concern when I started to recognize familiar landmarks.

I walked pass the Majestic Cinema, and when I got to the corner where the Kriloffs had their apartment, looking to the right, northward,

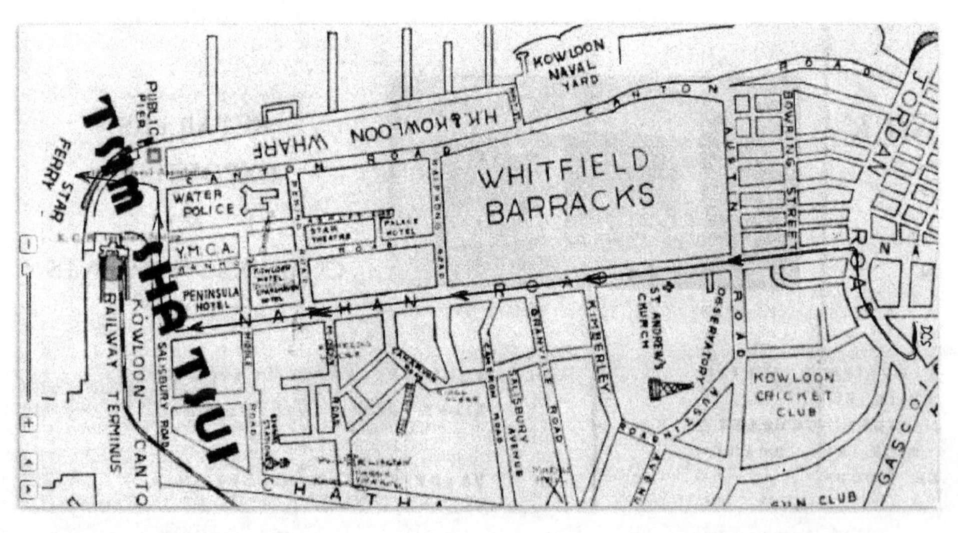

Courtesy Alison Ferry, gwulo.com

My escape route from DGS to the Star Ferry Terminal in Tsim Sha Tsui on Monday 8 December 1941.

I could see the Alhambra Theatre. Then, looking across Nathan Road, I recognized the building where Mother and I had lived so many years ago on the top floor. Then I set off on my trek southward.

I could see that Cherikoff's Bakery was closed when I reached Austin Road. A little further along, St. Andrew's Church and the church hall came into view. When I tried to gain access to the buildings, I found all the doors locked. Across the road and behind a row of trees on the right were the Whitfield Barracks, and on the same side but near Haiphong Road was the mosque, which seemed to have been there forever. At Middle Road, I recognized the Peninsula Hotel, and on the other side was Kowloon Park and Signal Hill, both very familiar to me. Around the next corner at the end of Salisbury Road, the Kowloon Canton Railway (KCR) Clock Tower came into view, and very nearby was the Star Ferry Terminal.

The Sewell Family's Unexpected Stay in Hong Kong

William Sewell and his family played a significant role in my life during this period, and because of this, I feel it is important that a little background information be provided to explain their presence in Hong Kong. Never could they have visualized that their brief and ill-fated appearance in Hong Kong would result in extreme privation for the entire family that would last for over three and a half years.

The following story was gathered from members of the Sewell family and research for this book:

William Sewell spent many years in China, mostly in Chengdu in Sichuan, where he eventually became the Head and Associate Dean of the College of Science. He was in Hong Kong to meet his family, who were to arrive from California on the American President Lines ship President Grant. Further travel arrangements had been finalized for the entire family to depart Kai Tak in Kowloon on the evening of December 8 for Chengdu, with a brief stopover in Chungking.

When their ship arrived in Manila, his wife, Hilda, and the three children had to unexpectedly disembark and from there make their own way to Hong Kong. Knowing that her husband was waiting for the family's arrival in Hong Kong, she had made valiant efforts to book a flight on a Pan Am float plane, to no avail, until at the very last minute, just before take-off, there was just enough room for the three children and one adult—there was no room for any baggage. All their luggage would have to be forwarded separately to follow them to Chungking.

William Sewell was totally surprised when his family walked in on him late on Sunday, December 7 at the Church Guest House in the central district of Hong Kong. When the excitement of greeting each other settled down, the explanation as to what happened to them in Manila was fully discussed. William also assured his family that all had been taken care of for their on-going flight the next day to Chungking, and not having their luggage was only a temporary inconvenience. The blessing was that now they were all together.

The next morning (December 8) during their breakfast they could see in the distance that Kai Tak was experiencing some unusual activities. As the family was scheduled to fly out of Hong Kong early that evening, he must have thought he should check with the airlines regarding their scheduled departure. This was particularly important as his family had a Hong Kong landing permit good for only 48 hours.

Having failed to receive any response from the airlines since yesterday evening, William Sewell must have decided that a quick visit to the airline offices at the Peninsula Hotel might be more effective and rewarding than waiting for them to contact him. Assuring his family that he wouldn't be away for long, he set out to make the 15-minute harbour crossing on one of the Star ferries.

Unfortunately, when he arrived at the Peninsula Hotel, he found that the airline offices were not yet open that early in the morning; when he inquired about taxi transportation to the airport, the police at the ferry terminal informed him that all traffic to the airport was forbidden because the airport was presently under a bombing attack.

Realizing that Hong Kong was under attack, heightened his concerns. To risk being stranded in Kowloon and separated from his family was sufficient reason for him to decide to return to the Church Guest House immediately and wait for further developments during the day. As fate would have it, these were the circumstances that led to the chance meeting between William Sewell and me on the same ferry early that morning.

My Encounter with William Sewell

Police personnel at the ferry terminal did not seem to have any control over the gathering crowds. One crowded ferry had just departed and coming up not far away was the next ferry. Suddenly apprehensive,

Courtesy PanAm Historical Foundation, gwulo.com

This is a picture of the Myrtle, aka the PanAm Hong Kong Clipper, which was on scheduled flights between Hong Kong and Manila in 1941. Mrs. Sewell and three of her children flew into Hong Kong from Manila on the Myrtle arriving into Kai Tak on Sunday, December 7th, 1941. Mr. Sewell had come from the interior of China to meet his family in Hong Kong and to return to free China with them the next day. c1940s.

gwulo.com

Arriving into Kai Tak the Sewell family took ground transportation to the airline's office (shown above) in the Peninsula Hotel on December 7, 1941. They then proceeded to cross the harbour by the Star Ferry to seek out Mr. Sewell, waiting for them probably at the Church Guest House on Upper Albert Road in HK Central. c1940s

I became increasingly unsure about what I was going to do next. Going back to DGS against a crowd flowing south would be almost impossible. My dilemma was soon to be resolved by outside forces.

The incoming ferry docked in its berth at the terminal station, and even before the disembarking passengers had cleared, the waiting passengers were scrambling to get on. The crowd surged, pushing me from behind, and I was carried forward, aboard the ferry and onto the upper deck, breathless and jostled forcibly against the starboard railing. I was small, not especially visible, so I clung tightly to the railing stanchions so as not to be trampled upon.

Amid the pushing and shouting about things I didn't understand, I looked around at the mixed assortment of passengers, Chinese and Caucasians. I was wearing a sleeveless woollen sweater on top of a light flannel long-sleeved shirt, with a green tie. Shorts and knee-high grey stockings and leather shoes completed my outfit. There were older students in different school uniforms, some carrying satchels or rattan school baskets; men and women in business attire going to work; a few children with their parents. Some of the passengers were carrying hand luggage. Everybody was talking excitedly with their closest neighbours; I wasn't aware about what!

When the overloaded ferry cast off to cross the harbour, it had a decided list to starboard—the side I happened to be on. I noticed nothing unusual in the harbour during the crossing, and I was unaware of any military activity; Green Island and Stonecutters Island were visible in the distance. Harbour traffic seemed normal and unhurried, but this would change drastically before the end of the day as Japanese forces overtook all of Kowloon before the end of the week.

I don't remember what prompted William Sewell to take an interest in me. There he was, standing quite close by at the time, and he must have been curious to see a small boy all by himself on the ferry, perhaps looking lost or even bewildered. The following could have been the conversation exchange between us:

"What are you doing here, boy?"

"I don't know, sir," I replied.

"Where are your parents?"

"I don't have any parents," I said.

William Sewell drew closer as he looked thoughtfully at me.

"What's your name, son?" he asked gently.

Encouraged by his friendliness, I responded, "My mother called me Robbie, but you can call me Robert; that's what everybody calls me now."

By then, the ferry was arriving at the terminal in Central, and the crowd was starting to get agitated in preparation for disembarking.

William Sewell then said, "You better follow me, Robert, as this is no time or place for a small boy to be left on his own."

We got off the ferry together at Central, and I followed him to the Church Guest House located at Upper Albert Road, not far from Bishop's House, where his family was waiting for him. Of course, they were surprised to see me in tow, but after he explained the circumstances to his wife Hilda, I was warmly accepted. This is how I met his wife, his two daughters, Ruth (8) and Daphne (6), and his son, Roger (4). They were all friendly towards me, and this made me feel comfortable and safe.

The Sewell family's greatest concern at this time was to be able to fly out on their scheduled flight this evening to Chungking, and all they could do was to wait for any confirmation from the airline agency. Because of the bombardment at Kai Tak today, nobody knew when it would become operational for civilian flights, but hope was still alive.

Because all members of the family had arrived in Hong Kong without luggage, they set off later that morning to do some shopping in the Central District. They went to one of the few shops that was open to buy some warm underclothes for the children, some slippers and some knitting wool. Without realizing it, this would be the only shopping that they would do for the next three and a half years. I too would be facing the same shortages over the same period.

That afternoon, the Sewell children were chattering to me about the passenger ship they were on from California to Manila (the President Grant). Then, about their flight from Manila to Hong Kong on the Pan Am clipper yesterday, Roger kept saying, "The big plane "brung" us." They also told me that they were going to leave Hong Kong that evening for Chungking. I don't know how much effort William Sewell made during the day to find out about their flight, but I know that he and Mrs. Sewell showed some anxiety about it. How did I know this? They were constantly reassuring their three children not to worry and that everything was going to be all right.

My greatest fear was what would happen to me if they departed that evening. Would I be left on the sidewalk again? I had no money to get on the ferry; how would I find my way back to DGS in Kowloon all by myself? Would it be opened? Where else could I go? All these thoughts and others crowded my mind. I thought naively maybe there was a possibility that if the Sewell family was to go to Kai Tak that evening, they could take me with them across the harbour and drop me off at DGS? That was a plan that wouldn't materialize.

§

Including the initial air raid on Kai Tak, there were two more raids during the afternoon. William Sewell insisted that we all take shelter at each subsequent air raid, all the time hoping against hope that the summons to the airport would come; alas, it was not to be. That night we all slept with our clothes on.

History has recorded that on this night, a total of 275 CNAC (China National Aviation Corporation) personnel, which included their families and the two Soong sisters and other VIPs were airlifted out of Kai Tak in Kowloon and flew to safety in Mainland China. The two Soong sisters were Soong Ai-ling, married to the richest man in China, H. H. Kung, and Soong Ching-ling, widowed wife of Sun Yat-sen, first president of the Republic of China. It was vitally important that the two Soong sisters not be captured by the Japanese. The planes were two DC3s and one DC2, all of which were in a hangar at Kai Tak. With such VIPs, the military must have commandeered the five seats originally booked by the Sewell family, effectively stranding them in Hong Kong.

This is how the first long day ended for all of us. The Sewell family had missed their flight to China through no fault of their own; they ended up without any luggage; worse still was the uncertainty of their predicament. As for me, I did not understand the gravity of my predicament; I had no doubts that the enemy was going to be beaten in combat, and I derived a lot of comfort in that thought and easily fell asleep that night. It was probably not so for the adults.

December 9

Last Hope for the Sewell Family Dashed!

Tuesday morning, we were woken early by the first of seven air raids that day, starting at 7:00 a.m. and lasting up to 1:00 p.m. The all-clear sounded at 2:00 p.m. As most of the bombing was confined to Aberdeen on the other side of the island, it had little effect on us nearer the Central District. Across the harbour, there seemed to be constant activity of one kind or another. Sporadic explosions were heard quite regularly, some of which may have been due to exploding demolition charges. I didn't know the difference.

William Sewell was listening intently all day to the news on the radio and looking out over the harbour and into Kowloon. He was probably still pre-occupied with the flight they had to miss from Kai Tak the night before and might have been wondering if there might be another flight that evening. Later that night, at about 10:00 p.m., once again we could see Kai Tak in the distance lit up and the noise from aircraft engines was heard. Apparently, there was an additional flight going out, but it was to be the last one. As disappointing as this was to the Sewell family, who by now were resigned to their fate, that night we were all thankful to enjoy a restful sleep.

December 10

Wednesday was going to be a horrific day (and night), much of which would be spent sheltering in the air raid tunnel underneath Government House. William made brief excursions between the shelter and the house between bombing raids and shelling to get food and other supplies as needed. I think this was the first time that I started to feel real fear creeping into my life! There were four air raids, two of which were in Central, where we were located.

The air raid shelter was packed with refugees of all sorts, civilian and military personnel. The ventilation system inside the shelter was so poor, breathing was only possible if you laid down on a blanket, so your face was close to the cold cement floor, which we did. Those who could, slept, and the rest endured the whole night. It was a great relief to get out into the fresh air the next morning.

December 11

On Thursday, air bombing was slight but shelling from the big guns from Kowloon was increasing. Once again, William Sewell shepherded all of us underneath the shelter at Government House, where we took shelter for most of the day, coming out once in a while when it was safe to do so.

December 12

Moved to the Peak

Friday, the bombing and shelling was the worst for Hong Kong. The Central District, where we were, was getting badly pounded.

Kowloon was now almost completely occupied by the Japanese army, who were shooting across to the island at will. All day long, the shells were screaming over us, and because of this, the Sewell family and I spent most of the day in the stifling underground shelter at Government House. It was fortunate that we were spared direct hits, but we did feel the shock waves from exploding ordnance every now and then when we were scurrying between the shelter and the house. I had a morbid fear of being separated from the Sewells from time to time, either in the crowded air raid shelter or dashing for cover on the street.

At the request of Dr. Kirk, who was working at the War Memorial Hospital on Mount Kellett, later that afternoon, Helen Kennedy-Skipton came to the Church Guest House to take the Sewell family to his home on Mount Cameron, as well as a Mrs. Raymond (62). There were some discussions about me, at the end of which I was relieved that it was decided that I would be included in the party. So, the seven of us crowded into the car and ended up on Mount Cameron at Dr. Kirk's house on 26 Middle Gap Road (aka 564 The Peak). In the house was a Mrs. Laird (61), who was recovering from surgery, and her husband, Mr. Laird (61). I do not remember seeing Dr. Kirk at the time.

William Sewell wrote in *Strange Harmony* that on this night his family slept on camp beds in the garage. I very probably was with them because of my friendship with Ruth, Daphne and Roger.

§

The following is an extract from a report written about events on Friday, December 12, by Sally (Sara) Refo when she was being repatriated to the USA on the Gripsholm, August 1, 1942:

> *"Dr. Kirk phoned Helen from the War Memorial Hospital on Mount Kellett and had asked her if she could come to the hospital for Mrs. Laird (who was recovering from a serious operation) and take her to his house on Mount Cameron, which was near Helen's house, and help her and her husband get food. Helen agreed. That afternoon Dr. Kirk phoned again and asked her to go down town central and get the Sewell family of five and Mrs. Raymond to add to the 2 Lairds in his house. She came back with one extra, an orphan child who had been in the Diocesan School and really had nobody to take care of him. "*

That extra person was me, the orphan from DGS!

Courtesy Alex Cooper

Dr. Kirk's house at 564 The Peak. On Friday December 12, 1941, the Sewell family and I stayed over here for a limited time before moving to 565 The Peak, to join the group at the Kennedy-Skipton house. Image: pre-war

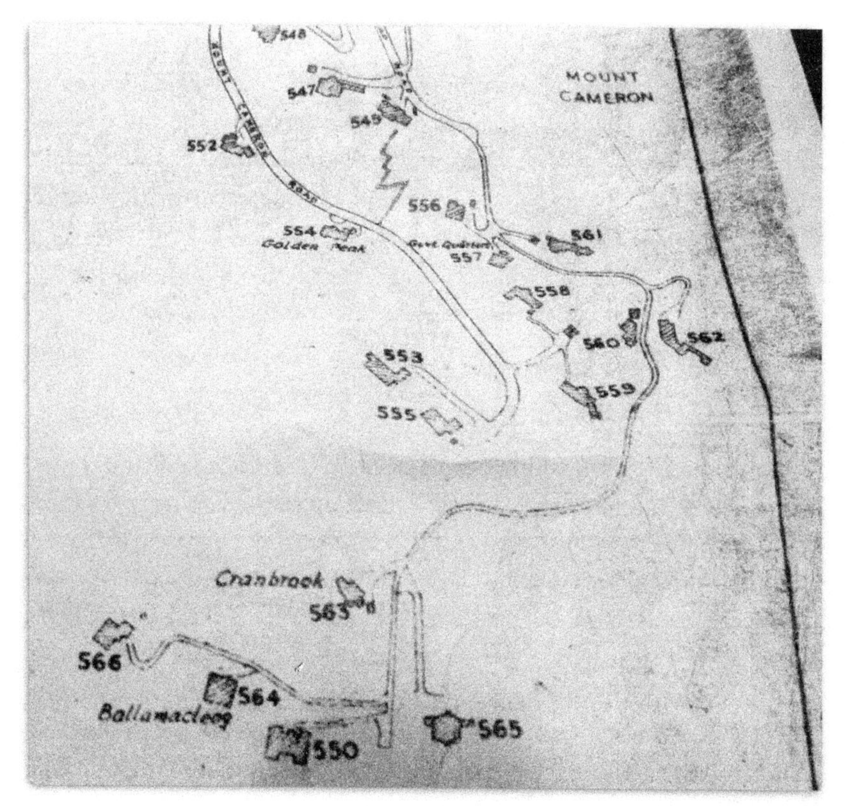

Philip Cracknell

Map of housing at Mount Cameron. Number 564 was Dr. Kirk's house, and number 565 was the Kennedy-Skipton's house. c1940s

After dropping us off, Helen then went home, which was practically next door at 19 Middle Gap Road (aka 565 The Peak), to join the group of refugees sheltering in her house. We were to join this group for supper later that evening.

In the Kennedy-Skipton house at this time were: George and Helen Kennedy-Skipton, their children Laetitia and Enid, plus two servants and their families; Marion Dudley and her servant with her sister; Ruth Milliken; Gladys Lynn and her daughter Ann; Henry and Jane Brownell and their daughter Betty Jane; Henry and Sally Refo, their children Alice-Ann, Peggy, Harriet, Burney, and two servants; and Miss Shinn, plus her adopted daughter.

That night, the two households got together for dinner at the Kennedy-Skiptons' home. Because of the large number of bodies, dinner had to be served in two sittings. I have no idea what Helen and Sally, with the help of their servants, cooked up for dinner for so many of us, but we were probably ready and thankful for whatever was served.

§

I do not remember how long the Sewell family and I stayed in Dr. Kirk's home. All I recall are the interactions with the children, all of whom were staying at the Kennedy-Skipton home. The move might have taken place the next day, or after the group returned to Mount Cameron after Christmas. For all the time that I was with this group on the Peak, I only have memories of the Kennedy-Skipton family, Refo family, Sewell family, Brownell family and the Lynn family.

December 13-21

At Mount Cameron

On Saturday morning, we had a good look at our surroundings. I remember well the long driveway leading up to the Kennedy-Skipton house. The house was modern and spacious and set in a beautiful open area almost devoid of trees and heavy shrubs. It could not be seen from City Central or Kowloon, and it also had a commanding view of the sea south of Hong Kong, the islands in the distance and the town of Aberdeen directly below us.

The next week was somewhat quiet, as the battle hadn't reached Mount Cameron. All the children in this group got along with each other. The younger children were able to play and explore the beautiful and still-peaceful subdivision, never far from the watchful eye of an adult. Some of the houses in the neighbourhood were occupied by servants of absent owners. This freedom of wandering around appealed to me immensely, knowing that I was safe and in care.

The American children all knew each other well, but they were friendly towards me and the Sewell children, we who were classed as British. We were welcomed in all the children's daily activities: playing, eating, praying, singing and attending Bible readings. Mrs. Brownell and Miss Shim were also giving us lessons, and all us children were encouraged to start making small Christmas presents; mostly creating and colouring Christmas cards.

There was some competition among the children in collecting spent ammunition shells, shrapnel and Japanese propaganda leaflets dropped from airplanes. We also agreed that war was much better than school, not realizing the seriousness of our situation. However, when there were air raids, we all had to take shelter in the basement of the house.

It was sometime during this period that the electric power was interrupted, which caused problems for the grownups to cook meals; they had to use firewood in Chinese cookers. Water too was cut off, and it was a bother to find and carry water and take care not to waste any. It was in this period that the population in the house started to be reduced; mostly the Chinese who departed to join their relatives in the city. There were some problems with the Chinese servants also, as Helen had let go her gardener and his family. I think Sally Refo retained one of her own servants. Maybe it was then, with more space available, that we became one group under one roof.

When food supplies were getting more difficult to obtain, those of the children who were able helped the adults scrounge for vegetables from the gardens of nearby houses. This could be heads of lettuce, cabbage, carrots, potatoes and even celery, which I remember I had difficulty growing at DGS. There were a lot of sweet peas growing too, and I enjoyed eating these peas as I picked them.

§

Towards the end of this period, we children could sense that the adults were becoming concerned about the outcome of the battle with the Japanese. We all became acutely aware that the front line of the Battle for Hong Kong was drawing closer every day to our idyllic location in the shadow of Mount Cameron. We could see dense smoke in the sky and burning fuel installation at Aberdeen. The reverberations from continuous shell fire were closing in, and aircraft bombing intensified throughout the day and into the night. Rainfall created havoc on some of the roads, impeding the movement of military wheeled vehicles.

I don't remember seeing any significant fixed military targets in our area, apart from movement of military equipment and men on nearby roads. However, a white building across the valley at one of the Gaps was visibly under shelling bombardment. It might have been a military command post at the time, but I never saw any direct hits. A short while later, I saw smoke seeping out of the shuttered windows of the building, which might have indicated that the building was set on fire. Maybe that was demolition?

December 22

Battle Intensifying on Mount Cameron

On Monday we started seeing increasing patrols of the Winnipeg Grenadiers and the Royal Scots. From the beginning, they were welcomed to rest in the house, and we children admiringly gathered around them, particularly the boys, examining their weapons and equipment as they amused us and gave us chocolate. Sometimes the soldiers appeared totally exhausted and the adults would give them a hot drink and some food, even a meal. One day, two of them were so exhausted that they slept for most of the morning before slipping away to rejoin their fighting unit. Others would only stay long enough to replenish their water bottles and immediately slip away into the surrounding hillside.

At this time, Mount Cameron was in the front line of the battle between our soldiers and the Japanese. Often in the dark, when we heard footsteps coming to the house, we were not sure whether we would see a Japanese patrol or one of ours. Outside on the lawn, we could see some of our soldiers digging trenches for defensive purposes. Nobody in our house undressed for bed that night, not knowing what was going to happen. We children sheltered in the basement and managed to get some fitful sleep, despite the frequent explosions of shells and grenades, the rattle of machine-gun fire and the steady movement of troops.

December 23

Escape to Mount Kellett

Tuesday, shortly after midnight, while most of us were still sleeping, I think it was a Canadian officer who appeared at the front door, urging the grownups to immediately prepare to evacuate the house as the Japanese troops were expected to overrun the area at any moment; he would be back shortly with transportation for us.

Many struggled into some warmer clothing in the light of flashlights, and still sleepy, we clambered up from the basement to the front doorway. I can remember it was dark and chilly outside, and not far away, we could hear sounds of machine-gun fire and muffled shell explosions. It was scary! Wanchai Gap was also lit up by some burning equipment.

As we began clambering onto the truck, a terrific nearby explosion rocked all of us, shattering some window glass in the house. Those of us outside fell on the ground for cover; meanwhile, the soldiers around us intensified their efforts to get us moving into the truck. Several of the children were so frightened they just lay on the ground, waiting for their parents to pick them up. In the darkness, I was scared and couldn't stop shivering. Anxiously looking around, I found the Sewell family comfortingly close by.

The explosion seemed to paralyze the adults momentarily, and in the mayhem, they were debating whether to flee with the soldiers or remain in the house and take our chances with the enemy; but the officer settled this. "We cannot hold the Japanese and are abandoning Mount Cameron," he said. "The Commanding Officer said you must leave for the Peak immediately." The adults dared not disagree.

By the time this decision had been made, the military truck assigned for us had been commandeered and driven away. Our disorganized group of children and agitated adults had no option but to set out on foot as we headed towards the Gaps in the darkness, trying to keep up with the flow of soldiers.

The moon shone faintly between moving clouds as we stumbled along the rubble-strewn road, mixing in with the mass of moving soldiers. The house to the left of ours seemed to be burning, and the flames were casting grotesque shadows around us as we passed it on our way down the road. To avoid enemy fire directed at Wanchai Gap, the soldiers led us into the bush, where we stumbled along until we emerged back onto the road further along towards Magazine Gap. Roger, being carried by his father, lost each of his shoes in turn, both retrieved in the dark, once by a soldier and the second time by his dad. Ruth Sewell tangled with some barbed wire and got a few gashes on her legs.

Daphne was giving her parents a fit by streaking ahead of everybody in the darkness.

A soldier saw that I was nervous about going off the road into the scrub, and that I was having difficulty keeping up with the Sewell family in the semi-darkness. He picked me up and carried me piggy-back until we got back onto the road. He was already loaded with rifle, bayonet, ammunition, kit bag, water bottle and so on, and my weight scarcely slowed him down. When he put me down, he turned around and pulled a prayer book from his pocket, which he handed over to me saying, "Keep this," and with a smile on his face, he vanished into the night to join his comrades; I never saw him again. I have this prayer book in my possession to this day, and on the front page is written in green ink: J.P. Martin, 14411799 PTE, Signals, 2 ESSEX.

Picking our way past Magazine Gap, the soldiers went one way, and our party regrouped to continue towards the Peak. By then, Daphne was securely in the care of her mother, while nearby, Ruth was trudging along with her father and Roger. In the clearing, the rest of us regrouped as we prepared for our trek towards our next destination led by Henry Refo; where it was, I had no idea at the time. With encouragement from the adults, we kids gamely trudged along on the dark road toward the Peak.

Just then, George Kennedy-Skipton came up from behind us in his car, having cleared a way through the troop congestion. He knew of a place ahead of us, near Mount Kellett, where we could find shelter and rest up for the rest of the night. He took the littlest children with him in the car and Henry Refo continued leading the rest of us on foot as we trudged along the road.

Quite exhausted, we arrived at the house where Henry Refo had settled the young children at about four o'clock on Tuesday morning. Inside, we all sank onto the cold floor to rest and try to get some sleep. Families snuggled together to share warmth. I had nobody to huddle with.

We were woken up at daybreak by the adults who were determined to find better accommodation along this road; it might have been somewhere along Kellett Road. Before long, we moved into a larger house in the shadow of Mount Kellett, where there was plenty of room for all of us. However, we had only the clothing we had on, and no food, water or bedding.

In the meantime, Mr. and Mrs. Brownell had wandered off and located a Canadian Army food depot (which was nearer to us here than when we were on Mount Cameron), and brought back milk, butter, bread and jam. Everyone felt much better after a good breakfast, and after last night's ordeal, we children were more than ready to explore our new surroundings.

I think the house we were in was occupied by a Dutch family living on the Peak, shared with another family. They probably looked upon us as invaders, but they did not bother our group. We found some bedding, dishes and a good garden with some vegetables growing. We made ourselves as comfortable as possible.

During the day, there were intermittent arcs of shells screaming over the house and landing on targets below in Aberdeen Town. We were safe from the shells if we could hear their flight path; the bad ones you don't hear because they land on you! At intervals, there were also air raids, but the area around the Peak appeared devoid of significant troop movements. We children were allowed a little more freedom to roam near the house and in the gardens, collecting shrapnel, which was still our favourite pastime. We couldn't hear the air raid sirens in downtown Hong Kong Central from where we were, so when the bombers were sighted flying overhead, the adults hurriedly gathered the children to take shelter. The other inconvenience we had to endure was a shortage of water, which had to be carefully rationed.

Because of the absence of law and order on the Peak, the greatest concern for absent property owners was looters. Probably with this in mind, George Kennedy-Skipton, with Henry Refo, drove his car back to Mount Cameron later in the day to see what had happened the previous night, if anything. We were all so relieved to see them return safely several hours later, this time with two cars laden with food stuffs and personal items left behind in the mad exodus a couple nights ago. They excitedly told us that they had safely escaped harm when an attacking airplane shot at them when they were on the road. That night, we all went to bed feeling calmer than we had thought possible earlier in the day.

Wednesday, December 24

Christmas Eve, we found ourselves looking down on an inferno not very far away. From burning oil storage tanks, below in Aberdeen, came volumes of dense black smoke. Other black clouds rolled in from the north. Air raids and shelling continued unabated, which caused us to frequently seek shelter in the garden grotto throughout the day.

The adults spent most of this day with the younger children, reading Christmas stories from the Bible. The teenagers were especially good with amusing the rest of the children. I recall how nice and full of fun Betty-Jane Brownell was to all of us; we called her BJ.

During the turmoil and destruction that night, all of us gathered close together on the floor around candlelight, raising our voices and singing Christmas carols. I had never experienced anything so sublime in my life, and it brought tears to my eyes. For an intense period, I felt a special kinship to all those around me, an experience I've never had before, and will probably never experience again with the same intensity. It was magical! Even to this day, I find these memories to be very emotional! That night, gifts were placed under pillows of children by their parents and exchanged between their friends. I do not recall whether I received anything, but I didn't care; I was thankful for the special gift that had entered my spirit.

Thursday, December 25

Christmas morning began with tea and bread for breakfast. Adults discussed Hong Kong's refusal to meet Japanese demands for surrender. There was a feeling of uncertainty, which heightened to despair for the young lady with a sick husband and a small baby.

Miss Milliken gave a cookie from her bag to each of us children. Helen Kennedy-Skipton had prepared a small gift for each of the children. I have no recollection of my gift, but I understand that Ruth Sewell treasures the gift she received to this day.

With our Christmas midday meal, Helen Kennedy-Skipton served a delicious plum pudding enjoyed amid the whistling of shells and what appeared to be a bomb explosion not far away, which shook loose some plaster in the house. We abandoned the plan to play games in the garden and turned to reading some of the house's plentiful supply of books instead.

In the afternoon, enough water was collected from a nearby stream that everyone could have a skimpy Christmas bath. I had just finished my bath and was standing in the glass-enclosed porch looking out on the garden and combing my hair when a solitary airplane flying high overhead dropped a bomb, which landed in the garden before me. The effect was electrifying! Just a few seconds before I heard the explosion from the garden, I saw the entire porch window starting to cascade towards me. I just had time to scoot underneath some camp beds lined up against the far wall, and apart from being stunned by the explosion and sustaining some minor cuts from flying glass, I was unhurt. There was no other damage to the house, nor did any fire break out.

A little while later, we all gathered outside looking at the huge, still-smouldering crater in the garden. The adults talked among themselves as they surveyed the scene, and we children were gingerly picking up jagged steel splinters, some of which were still hot to the touch.

As the hum of the bomber receded, we became aware of a strange new stillness. Not a sound was heard. The strange hush was even frightening.

"We have surrendered," someone said. It was unbelievable!

Then, very quickly it seemed, the air started to fill with the song of birds. This lifted all our spirits, and the children out in the garden even became frisky, running around and playing tag with each other. Most of the time BJ was 'it' and laughingly chased us. Several times, I was caught up in her arms as she scooped me up off the ground; the feeling of relief in the air was intoxicating. The bomb that fell in the garden of the house on Mount Kellett might well have been the last one to fall on Hong Kong.

Soon after, we thoroughly enjoyed our Christmas dinner of stew and mincemeat, with brandy for the adults. In the meantime, one of the tenants in the building came to us and announced that Hong Kong had indeed surrendered this afternoon. There was great joy that we would now be spared any further bombardment, but together with this relief was concern of what might happen next. However, with this new realization, we felt more of the Christmas spirit than we had felt earlier in the day and were thankful that we had survived up to now.

That night, the men deliberately stayed awake to watch out for any Japanese movement in our area. The women stayed awake involuntarily. We could hear the cries of people and animals in Aberdeen far below, directly beneath us. The dairy farm was in Aberdeen, and it sounded like the looters were killing the pigs and dairy cows. The rougher elements of Hong Kong (some fifth columnists, some robbers, some desperately hungry people), as well as the Japanese, were all actively engaged in wanton looting. During the night, we could also hear the screams of women rising from Aberdeen. We did not know if the Japanese would come to our house or not. Our every breath was a prayer.

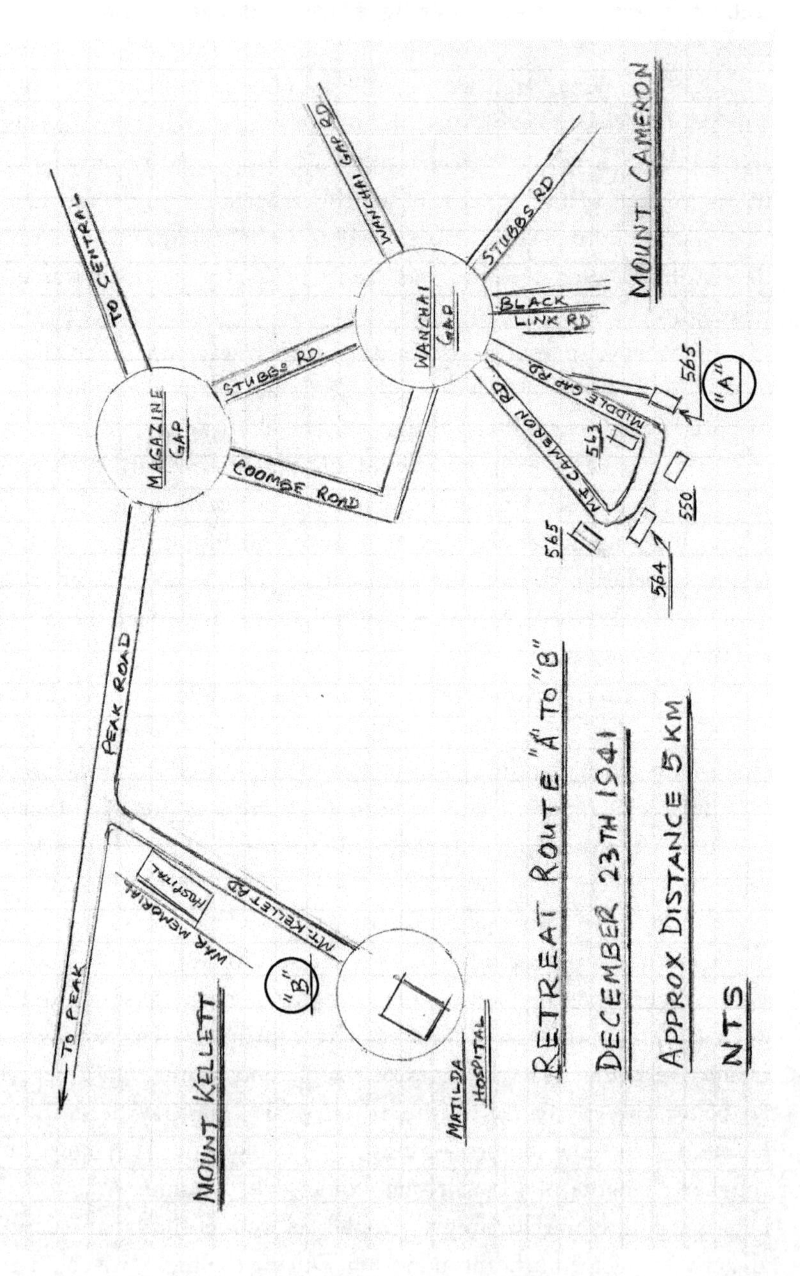

Escape map to Mount Keller

CHAPTER 7:
DAYS AFTER SURRENDER

Surviving Under Occupation

December 26, 1941 – February 11, 1942

There were two groups of civilians on the Peak the day after the surrender: our group was located on Mount Cameron in the Kennedy-Skipton's house. The other group was located on Mount Kellett, probably centred at the War Memorial Hospital near where our group was on Christmas Day.

Friday, December 26

Return to Mount Cameron and the Days that Follow

This was going to be an action-packed day. Straight after breakfast, we all gathered our belongings, helped load everything into the two cars, and were soon on our way back to Mount Cameron. Mount Cameron felt like home, and we were all looking forward to being back in familiar surroundings. Mrs. Kennedy-Skipton's concern about the vulnerability of her unprotected home to roaming looters was made more urgent by the lack of civil order immediately after the imposition of Japanese military control. The horrible sounds of screaming women emanating from Aberdeen down below last night was still haunting us and heightening our fears.

Mr. Kennedy-Skipton drove one car, and Mr. Refo drove the other. Women and children were crowded into the front seats, and some of the adults and older children walked beside the cars, carrying their belongings. We moved slowly and kept together.

The first sight that greeted our eyes were great stone buildings, war-torn, damaged and deserted. We slowly drove past the War

Memorial Hospital on our left, and moving along, crossed Magazine Gap without incident.

Continuing, we turned the corner at Wanchai Gap, and encountered a group of Canadian soldiers crouching on the side of the road, under guard by Japanese soldiers with rifles and fixed bayonets. The Japanese officer drew his revolver as he approached. We stopped, laid down what we were carrying, and crouched in the gutter by the side of the road.

The adults explained to the officer that we were trying to return to our home on Mount Cameron, not far away. After some discussions, Mr. Refo was made to unload his car and drive a Japanese soldier into town. We didn't know whether we would see him again. Some of the Japanese soldiers were friendly to the children while we sat by the roadside and waited. Finally, Mr. Refo returned in the car, and after some discussions, the officer wrote something on a piece of paper for us to show other Japanese soldiers if we ran into more difficulties and then waved us away.

Elated and without wasting a moment, the ragged procession reformed, and we moved away as quickly as possible. However, on rounding the next corner, the cars were stopped again by some unfriendly and armed Japanese soldiers. These soldiers looked at the note that the Japanese officer had written and only then allowed our men to walk up the long driveway to the house. The rest of us waited anxiously until our men returned a little while later. There were some Japanese occupying the house, and they refused to move out.

Mr. Refo and Mr. Kennedy-Skipton went back to explain the situation to the friendly officer. He advised us to wait until the next day, when the soldiers would most likely move out, and then we could move in with no problem. We did not know where to wait. It was midafternoon, and we were a few miles away from the house we had left earlier this morning. Some of the children were getting tired, hungry and thirsty.

There were four other houses near us: one was vacant and locked, another had been damaged by the fighting before Christmas, one was occupied by Norwegians and the other by a Swiss national called Mr. Ott. He agreed to give us shelter for the night and let us use his kitchen to prepare something to eat that evening. Afterwards, we put together some bedding on the floor and all of us settled down for a restful sleep.

Dr. Kirk's house was probably locked up or in the hands of servants, as no doubt, he too must have moved out of Mount Cameron before Christmas and taken up residency at his work station in the War Memorial Hospital on Mount Kellett.

By mid-afternoon the next day, there were signs that the Japanese were preparing to leave the Kennedy-Skipton house. When they finally did leave, the adults all went over to check things out. They found the interior of the house in a shocking state and Helen's amah a nervous wreck; she had been tied up in the basement while the Japanese quarrelled amongst themselves as they looted the house.

As disheartening as the state of the house was, the adults immediately set to cleaning it up. The children stayed at Mr. Ott's house for another night, and Mrs. Refo prepared an evening meal for all the hard workers. The cleaning wasn't completed until the next day, but by then, it was sufficiently habitable; it was home, and we were delighted to be back. Cleaning and getting the house back to a state of normal lasted several more days.

§

In the days that followed, we lived as much as possible in the garden and were almost entirely isolated from the few remaining fellow neighbours. We did not venture far for fear of running into one of the frequent Japanese patrols in our district. Mrs. Refo organized schooling for us children.

The days were growing colder. Someone found a case of Canadian hospital cotton smocks and trousers, and idle hands were put to work unpicking the seams of these items, so they could be sewn into trousers for the children and jackets for the women. This was particularly useful for the Sewells, who were all very short of clothing, and me, still wearing what I had from school.

As there was very little food left in the house, we had to start foraging to replenish supplies. One of the Chinese women (in our group) wrote a note in simple Chinese characters: "We have eighteen children; may we have some food?", and two more to show to sentries we might meet: "Please may I pass to buy food for my family?" and "Please allow me to go to my home, which is on this road." These notes proved invaluable. The first one helped when parties of Japanese soldiers were clearing away portable Canadian Army stores. They gave us several cases of evaporated milk, cheese, corned beef and vegetables, which we were all happy to bring back to our home where Helen Kennedy-Skipton stored and used them as seemed best.

The Japanese considered the Canadian Army stores their own, but we felt they were ours if we could get to them first. We knew we would be interned before long, so we gathered what we could, in preparation for the inevitable move. In between groups of Japanese picking over the stores, we gathered what we wanted—blankets, dishes, raincoats, wash basins, tools, cutlery, military clothing, stationery supplies, etc. I picked up a blanket and some T-shirts that were much too big except as nightwear, and a neat pair of goggles meant for motorcycle dispatch riders. We children set up a small market at the side of the garden to display the merchandise we had gathered.

The adults also went along the hillside to other areas of the Peak to collect more food from other military depots that were still open. They had to pass two lots of sentries, one in each Gap; these forays always involved a certain amount of fear, for those who went and for those who waited for their return. We were so happy and relieved to see them return safely and with a good quantity of supplies.

With time, we grew less and less dependent on these trips for food, and instead rifled through the empty houses close around us. While the Japanese ate their meals, we systematically searched the neighbouring vacant houses on Mount Cameron to discover any food we could rescue. There was special joy when we found two great big bags of potatoes, each weighing over two hundred pounds, but it was a struggle getting them back over nearly half a mile of road. We harvested lettuce and vegetables from the gardens, and once we even found a leg of ham in a cellar.

Soldiers on Our Doorstep

We were always liable to have Japanese appear on our doorstep. "Visitors, visitors," our lookout would cry, and the older girls would vanish to a little room at the top of the house. It was always Alice Ann, Betty-Jane and the young amah who had to be kept out of sight. The Japanese were known to be tolerant towards children, so the adults encouraged us children to cluster around the front door to greet them.

Mr. Refo was the main lookout person, but all the adults would take a turn at this essential duty, all for the protection of the young girls and ladies.

Sometimes the visitors were gendarmes, like the gestapo in Germany, coming to get our particulars: name, age, genders and nationality. Sometimes they were officers or uniformed men off duty, having a look around their newly conquered territory.

Our visitors usually demanded food, and to forestall them, we'd offer tea at once. This small courtesy turned their natural politeness into graciousness. They would play with the smaller children and amuse them for hours or play on the piano together with the older girls. Sometimes they went into the garden to practice shooting old tin cans with their revolvers. This particularly impressed me and Roger.

§

One morning, some Canadian soldiers convalescing in the War Memorial Hospital not far away walked down to Mount Cameron. We gave them breakfast in our kitchen while someone kept watch; fortunately, no Japanese came. They informed us that wounded soldiers were not imprisoned until they recovered from their wounds, so naturally, their convalescence was made to stretch as long as possible.

One day, two Japanese soldiers arrived at the house when only Mr. and Mrs. Refo, Mrs. Dudley, some of the older girls and a few of the children, including me, were at home. The soldiers opened the door and came in, revolvers in hand and shouting. We were all totally scared, not understanding a word they were shouting.

Seeing the Canadian Army mattresses on the floor, they said "Canadian soldiers," in an accusatory way.

"Children, many children," the Refos told them, again and again, but the soldiers shook their heads and searched the house. They soon found the older teenage girls, who had disappeared upstairs at the first sight of Japanese company.

The soldiers counted more beds than were people in the house. Because these were army-style beds, they were convinced there were soldiers somewhere in hiding. Just as things were getting heated, the rest of the party returned. Mrs. Kennedy-Skipton immediately offered the soldiers some chocolate, which appeased them, but it was matching the body count with the mattresses that convinced them we weren't harbouring any soldiers. Reassured, the two men spent the rest of the day teaching the children Japanese tunes on the piano and learning in return how to play the tune to "God Save the King", using one finger on the keyboard. It was quite funny!

Early one morning when it was still dark, Mr. Kennedy-Skipton found a blood-soaked Chinese man staggering up the long driveway leading to the house. He had stab wounds in his chest, side, back, hand and one calf, and one thigh was slashed open, revealing the bone from end to end. There seemed little chance of saving his life—his dirty clothes were stuck to his wounds and stiff with dried blood; he was pale, cold, weak and in pain.

At breakfast in the morning, we learned about this man and how badly he was wounded by the Japanese soldiers. Ann Lynn's mother (Gladys) and Mr. Refo were tending to him, and they both wore a grave look on their face, wondering if this poor man was going to survive or die from his injuries. After a few days, we were all delighted to learn that he was slowly recovering from his ordeal. When he was strong enough, he told us that the Japanese had killed his three companions and left him for dead. That's how he dragged himself to our house.

It was fortunate that frequent Japanese patrols never discovered him in the house. They came to see us nearly every day, interrupting everything we did and going where they pleased in the house. It took some tact and distraction to keep them from finding the wounded man or the older girls who disappeared upstairs when enemy soldiers came to call. As usual, we distracted the soldiers by giving them sweet peas from the garden and cigarettes. They liked to talk to the children, and loved to play the piano, or listen to Peggy Refo, Sally Refo's daughter, play for them.

The concern was to conceal him from the frequent Japanese visitors, as we did not know what their reaction would be if they found him. We were all told by the grownups how not to give away the secret about our wounded man. Whenever a Japanese patrol would stop by the house, all of us children would gather at the doorway and crowd around the soldiers, helping to distract from any signs of suspicion. Of course, the adults also helped in this distraction.

He stayed in the house for a few weeks and recovered steadily until he was well enough to leave us. He was so grateful to Mr. Refo and Gladys Lynn for their care, and to the amah that fed him, and to the whole house who protected him. Then one day, he quietly slipped away from the house to return to the town below, carefully avoiding being seen by any soldiers, in case we would get implicated in his rescue. We never saw him again. This was an amazing event, something I will never forget!

On another day, shortly after the above event, some of the girls came across the bodies of two Canadian soldiers on the hillside, about two hundred yards from the house. They were quite decomposed and had been dead for some time. The adults buried the bodies. No one felt like exploring after that, and the sight of any buzzards wheeling above the nearby hills made our hearts uneasy.

§

A few days after the wounded Chinese man left our house, an aircraft dropped leaflets over Mount Cameron. Naturally, all the children dashed out to pick them up. The message was that all British were to report to the Murray Parade Grounds in Central by a given date. The deadline had passed by the time the leaflets were dropped to us, so the adults decided that the orders did not apply to our group.

That afternoon, a gendarme officer who had inspected us before visited us again.

"All the British are segregated. Why are you still here?" he said to us.

The adults told him there were many children in our group and we needed transportation. There were 10 children, and some of the adults were not strong enough to walk far. Mr. Brownell volunteered to go to register on behalf of everybody in the group and request transportation at the same time. In the meantime, we all packed our bags and bedding rolls and waited for him. Mr. Brownell never came back, and we never saw any transportation.

At this time, the group destined for Stanley Camp consisted of the following: Mr. and Mrs. Refo and their four children; Mr. and Mrs. Brownell and their child; Mr. and Mrs. Sewell and their three children, plus me; Mrs. Lynn and her child; Mr. and Mrs. Laird; Mrs. Milliken; Mrs. Dudley.

Mrs. Brownell later learned that her husband didn't come back because he was seized by the gendarmes at the registration centre and sent directly to Stanley Camp on his own. Mr. and Mrs. Kennedy-Skipton and their two children were not interned, as they claimed neutral Irish nationality.

By this time, many of the Chinese ladies with children had already left the group, including a significant reduction of servant staff and their families. At the height of the emergency, Mr. and Mrs. Kennedy-Skipton had provided shelter and were feeding up to 40 bodies at one time! A truly remarkable couple.

Two days later, two staff cars drew up, and out stepped a major in smart uniform, with several subordinates in attendance. Their swords swung noisily, and their heavy leather boots clattered on the doorstep. (The following dialogue is by Mr. Sewell in his account, Strange Harmony).

"You must all gather," said the major, in English.

This was different from any previous encounter and we clustered in a tight group.

"You are all here? How many?"

"Twenty, ten are children."

"All British and Americans were ordered to muster and are segregated," said the major severely. "You have broken martial law."

He glared at us, but no one answered him. There was a sudden movement and a small voice broke the silence.

"I'm Roger, how are you?" said Roger Sewell to the major.

There was a roar of laughter. Four-year-old Roger Sewell shrank into his parents' arms, shamed to tears. But the tension was broken, and relief had come in its place.

"So, you are Roger," said the major, patting him on the head. "Who are all the others?"

We were then each introduced to the major. I don't recall what his reaction was when I was introduced without any parental backup. Maybe he thought I was related to the Sewell family.

The major seemed impressed and told us to wait for further instructions. The next day, he came back with two sacks of rice and other food and told us to wait for further instructions. A week later, he came back with more food, and the day after that, a Japanese guard brought a frightened Chinese photographer to the house. He photographed all of us in groups of two or three.

A few days later, all the adults were issued with a Japanese military pass allowing free movement in Hong Kong. Since I do not remember receiving a pass, perhaps the children were excluded. The adults were delighted with this 'freedom of movement', which made it easier for them to prepare for our eventual move to Stanley Camp.

§

We celebrated the arrival of the passes with a trip to the Peak, where we met another group of British civilians waiting for internment, like us. They had no passes, so were confined to their district and were only allowed access to a nearby army field food depot.

Once or twice a week, the major came, usually bringing candies or presents for the children. The smaller girls made bouquets for him, which he took away, his face wreathed in smiles.

"What do you want today, children?" he asked one day.

"My daddy needs a haircut," cheeky Daphne Sewell said.

Later in the day, a crestfallen Chinese arrived, saying he had instructions to cut everybody's hair. I am almost sure that I also had a haircut.

§

We all knew that before long, we should have to be interned, so we started to gather what we could in preparation for the inevitable move. At first, we had felt guilty as we ransacked our neighbours' unoccupied homes and rifled dumps of Canadian Army stores lying around in profusion. First one batch of Japanese would come and pick them over, then another would make his own selection. In between, we gathered what we wanted—raincoats, washbasins, tools and bowls. I joined some of the older children to run a 'stall' at the side of the garden to sell or barter smaller things amongst ourselves.

With the freedom of the passes, Sally Refo, with three of her older children and William Sewell, decided to visit the Peak again. When they arrived, they were greeted by quite an excited group, preparing to leave for Stanley. They watched several parties depart, packed tightly into trucks, while some trudged with a few possessions on their backs. The nearby field food depot had already been emptied of its contents. Soon another truck passed, with the last of military patients from the nearby War Memorial Hospital (WMH). In their midst was our friend, Dr. Kirk, who shouted from the truck, "If you want food, there is some in the hospital store. Help yourselves!"

In the hospital store, they found a treasure trove of all kinds of food stuffs, more than they could carry back to Mount Cameron, about two miles away: stacked tins of butter, sugar, treacle, candles and salt, as well as toiletry necessities of every kind.

§

The gendarmes were the only visitors we really did not like. They were always unpleasant and continually told us that we must go to Stanley Camp for internment. They said our passes made no difference, and we must go to Stanley. They scheduled a date for our move and said that they would come along with transportation to load us up on that date.

Meanwhile, Henry Refo and Mr. Skipton took the matter up with the Japanese army who had given us our passes and found out that the gendarmes were determined to send us to camp. However, they were agreeable that we leave the Peak to first move to Bowen Road, but permission must be obtained from the Army, Navy and the gendarmes.

The officer who came to see us next was the kindest we met during the whole war. He seemed sympathetic to the adults' concerns and told us to wait for further instructions. The next day, he came back and brought us children some food and asked more questions. After further discussions with the adults, a week later he came back with lots of food, including waffles for the children. I think it was at this time that final permission was given for us to move to the residence of the Refos on Bowen Road.

§

Mr. and Mrs. Refo had been down to their house on Bowen Road beforehand, to prepare it for our planned move. They had to clean up some debris caused by earlier fighting in the area, but fortunately, the house had not been looted, likely because the nearby house overlooking the harbour was being used by the Japanese Navy and was guarded by sentries. Their task was to guard the nearby waterworks.

Leaving Mount Cameron and the hospitable Kennedy-Skipton family was quite emotional for all of us. The Japanese allowed them to remain in their home on Mount Cameron because they were Irish, and for their own protection, they flew the Republican flag of Ireland in front of their house. This I remember quite distinctly.

Move to Bowen Road

Number 14 Bowen Road was the address where the Refos lived before the war, and they were eager to go back and gather things they would like to take to Stanley Camp. The truck supplied to move us to Bowen Road from Mount Cameron had to make several trips to transport all the stuff that the group had with them. I think most of this stuff belonged to the Americans, as the Sewell family and I had very few possessions apart from what we had gathered from our own military stores on the Peak. The day of our move to the Refo's home coincided also with the scheduled day that the gendarmes were coming to ship us to Stanley Camp. When these gendarmes arrived with their two trucks and saw what was going on, they left us alone and instead drove their trucks around the Peak district, apparently looting vacant homes of their furniture.

During my research, I discovered a most remarkable photograph. As far as I am aware, this is the first time that it has been published. The photograph was taken by elements of the Japanese Navy on the grounds of #14 Bowen Road. The photographer was probably taking this photograph for propaganda purposes as all of us were very well dressed for the occasion. I was still wearing what I had on at the beginning of the battle. The image depicts a group of children just prior to internment in Stanley Camp.

After we settled ourselves at Bowen Road, the gendarmes and army seemed to forget about us. But the naval officer in charge of the waterworks just below the Refo house was friendly and came to see us every day, often bringing sugar, butter, bread or candy. This officer, who we called 'the Admiral', had probably been told by the gendarmes to keep an eye on us, and he did his duty well. He seemed to like Mr. Refo, Mrs. Raymond and Mrs. Laird best and always called for them, but his long visits exhausted them. He didn't always endear himself either, such as when he told Mrs. Raymond, an Australian, that the Japanese would soon take Australia, and she could then go home.

By this time, William Sewell had expressed his readiness for internment. I didn't quite know what that meant, except that it might mean another move to somewhere. Would I be included in this move, staying together with the group? The Japanese must have assumed I belonged with the Sewell family and anyway, what could William Sewell have told the Japanese, or anybody, about me?

§

Sometime during the first week of February, a short note came from the army major, and it was the last we heard from him. It said: "Day after tomorrow, you will be interned in Stanley Camp."

On February 9, 1942, two trucks came to take us from Bowen Road to Stanley Internment Camp. Thus ended the saga on the Peak, and our captivity began.

It had been a stressful time for me, between the battle and the uncertainty of what was to become of me. But the adults had no idea either, as far as I could tell. I was not sure whether I felt sad at the prospect of being separated from the group with whom I had shared the intense ordeal of the fighting on the Peak. Perhaps I was too young to recognize whether any of these adults had any sympathies for me, now that my 'free ride' with the Sewell family was about to end.

I had spent 62 days with the Sewell family and the rest of this group on the Peak, and yet I have little memory of the development of any personal relationships. Sharing extreme adversities and common danger to life and limb under wartime conditions, one would think, would be a natural catalyst for the development of intimate and lifelong relationships. Perhaps I was too young to be aware of such emotions, or maybe I was in shock.

Betty-Jane Brownell had reached out to me several times and had been kind to me, and that left a lasting impression on me. I will remember the Kennedy-Skipton family with gratitude, and I will never forget William Sewell and his family.

Courtesy Ruth Sewall-Baker, Unknown Photographer

Group at Refo Residence on Bowen Road, c Jan. 1942.
1. William Sewell. 2. Ruth Sewell. 3. Daphne Sewell. 4. Roger Sewell.
5. Robert Tatz. 6. Betty-Jean Brownell. 7. Alice-Ann Refo.
8. Burney Refo. 9. Peggy Refo 10. Anne Lynn. 11. Harriet Refo.

CHAPTER 8: WARTIME INCARCERATION

Safety in Numbers

1942 – 1945

Common adversity is a great leveller of society. Pain, hunger and danger suddenly become an essential component of everyday life. The privileges and safety that wealth, power or status bring to some suddenly melt away with the heat of the battle. All of a sudden, we discovered that we are all more similar than not. This was acutely evident during the fighting and in Stanley Internment Camp.

Stanley Internment Camp – February to June, 1942

We looked around at our new world as our two trucks piled high with the 20 of us and our belongings came to a halt inside the camp boundary. Stanley Camp was located on a green hillside with a view of the brown, rocky seashore below. The surroundings were a pleasant surprise to many of our party, who were relieved and more content in the place than we had expected to be.

Groups of people watched us from their sunny seats on the grass or from the football pitch as we dismounted from our trucks, piled high with 20 people and our personal effects. Dismounting, we sorted our belongings and waited to be told where to go.

Our American friends were directed to the three blocks, A2, A3 and A4 on 'Roosevelt Avenue', allocated for Americans, which faced the communal kitchens and workshops. The Sewell family and I were directed to the nearby Married Quarters: Blocks 2, 3 and 4 attached to each other, and Block 5 stand-alone, all sharing a common central courtyard. These were all staff quarters originally occupied by prison

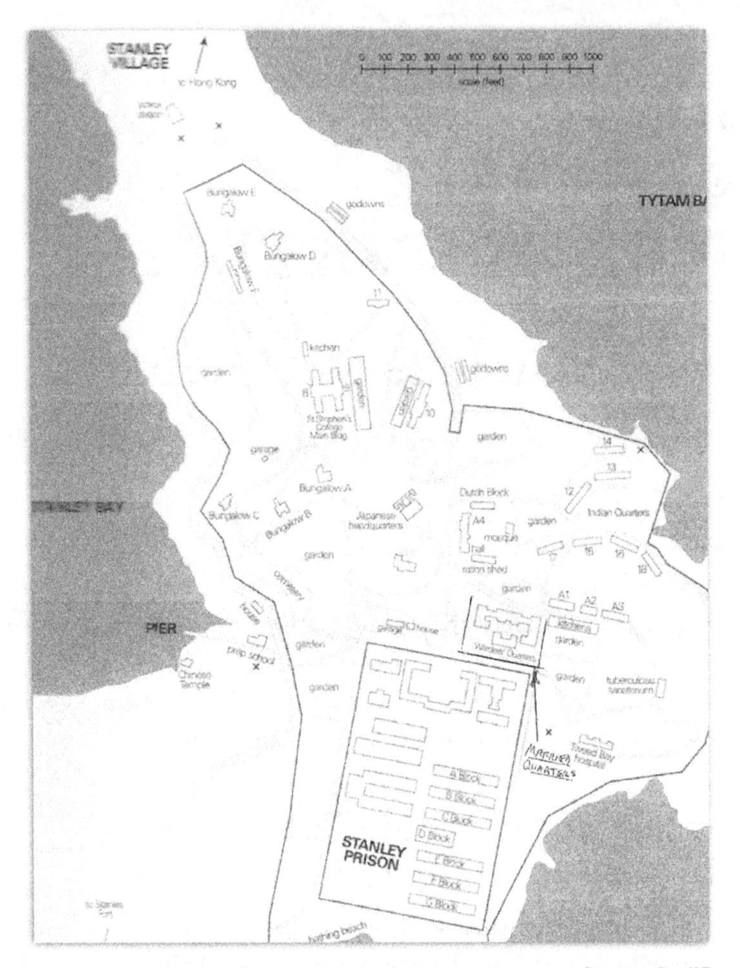

Courtesy Geoff Emerson

Plan of Stanley Camp Complex and Married Quarters. c1942

officers and their families. The Sewell family was billeted in Block 5, on the southwest corner overlooking the prison wall and facing the Japanese camp HQ on the west. I was assigned space in the middle of the same block 5, but in Room 17, which faced south over the prison wall.

The event I had thought inevitable and dreaded now occurred; the Sewell family and I were separated. Mr. Sewell must have answered honestly when the billeting officer asked about my relationship with his family, and I soon found myself standing alone again, as I had on the morning of Monday, December 8, 1941, on the pavement outside DGS. Again the question arose, who is this orphan boy?

Image by photographer on HMCS Ontario, *forposterityssake.ca. Reference DND/DHP Annotated by Philip Cracknell.*

Stanley Camp Prison Officers Club & distant Windy Gap. c1945.

I cannot recall the exact configuration of Room 17, just that it was fully occupied and there were eight beds with stuff piled on them in my room. I never explored the other rooms. With no acknowledgement or help from any of the adults in the rooms, I dropped my Canadian Army bedroll on the floor near the entrance to the large living room. I added my other meagre possessions—an army mess tin and utensils, a blanket, army goggles, empty brass bullet casings, a worn eraser, a few rubber bands, some clothing items and other sundries and went out again to have a closer look at my new surroundings.

§

The greatest difference between me and the other orphans at Stanley Camp was that I had no credentials or mentors. Miss Gibbins, the top official at DGS, didn't provide me with any identification documents before leaving me alone on the street at the outbreak of the war. When I arrived at Stanley Camp on February 9, 1942, I had explained my story so many times my answers were nearly pat:

"My name is Robert Tatz. I was born in Hong Kong. I am an orphan. I lived at DGS in Kowloon. I have two older sisters at a convent school, but I do not know where it is. I was lost in the streets of

Image by photographer on HMCS Ontario, *forposterityssake.ca. Reference DND/DHP*

View of Blocks 2 and 3. Maximum Security Prison in the background under control of the Japanese Military. Stanley Camp Married Quarters. c1945

Kowloon the day the war started. Mr. Sewell and his family cared for me during the Battle for Hong Kong. I arrived in camp with the Sewell family."

This was the only information I could have given at the time when I was interviewed. There was no question that my circumstances were unusual; it even baffled historians researching this period many years later.

Surely, out of about 3,000 residents in Stanley Camp, there had to be somebody who would recognize me, or know of me and my mother.

Subsequent research revealed that my godmother, Babushka and Uncle Harry were all in camp. Miss Gibbins and Reverend Dean Rose, both from DGS and St. Andrews Church, were also in camp. Mrs. Rodgers, the kind American woman, was also there, as were Aunty Lydia's brother Vitaly Veriga and Dr. Uttley, who had driven me up to stay with my mother at Matilda Hospital. Any one of these people could have sorted out the lingering, troublesome issue of my lack of a verifiable identity. They are all long dead now, so the question of why none of them stepped up to do so will remain unanswered.

It was paradoxical that, at 10 years of age, I would find life in the internment camp a joyful experience. And why not? I was to endure no adult discipline; nobody chased me to brush my teeth; I didn't have to comb my hair (I had no comb or hairbrush anyway); I was thankful that nobody took an interest in my threadbare attire—my ignorance of acceptable standards in cleanliness and hygiene was nobody's business—and attending available limited educational classes was entirely optional. I remember well that the shower water was very cold, but I have no memory of what I used for soap, or if I even had any.

As my residency in Stanley Camp was limited, my memories from this period do not reflect anywhere near what others experienced during the three years and eight months of their incarceration (January 1942 – August 15, 1945); that's their story! To learn a more complete story of Stanley Internment Camp, the interested reader can refer to an excellent book by Geoffrey Charles Emerson, entitled *Hong Kong Internment, 1942-1945: Life in the Japanese Civilian Camp at Stanley.* With diligent research, a reader will discover many more accounts of personal survival describing their own unique experiences. Here, I recount what, at 10, I witnessed, and how I tried to make sense of a world created by war, so different to what I had experienced in my short life.

The general feeling among the internees was that their current adversities would be short-lived. The inevitability of the Allies' victory was never in doubt, though the strength of this belief would be sorely tested by anxieties and dashed hopes as the weary days, weeks and months dragged on into years, in which the war seemed unlikely ever to end.

The internees' belief in the Allies' inevitable victory was also shared by the local Chinese population living in Hong Kong, especially former employees of commercial enterprises and civil servants. With great ingenuity and risk to their lives and safety, these kind people sent food parcels into camp, smuggled in cash and extended credit to camp inmates, all under the nose of the feared gendarmerie. However, history has recorded severe reprisals, arrests, betrayals, tortures and even executions, to those caught in such activities.

Presumably, in my circumstances at the time, I would have fallen under the welfare and billeting committees. Why welfare? I was a juvenile orphan with no mentorship or sponsorship, who desperately needed support and protection. Why billeting? As a vulnerable juvenile, I needed safe accommodation in a safe environment.

Stanley Internment Camp was located on the Stanley Peninsula southeast of the island, about 10 kms from the city centre, on grounds adjacent to the maximum-security Stanley Prison, which was not part of the camp grounds. The prison itself was enclosed by a massive 18-foot wall, and the grounds outside this wall had housing for over 800 staff and officers. The Camp also encompassed the grounds of St. Stephen's College and Preparatory School, which included accommodation for college staff and student boarders. The nearby Colonial Military Cemetery was also part of the camp grounds. This whole area was enclosed with barbed wire fencing, and subject to regular Japanese patrols. It was Dr. Selwyn Selwyn-Clarke, the Director of Medical Services, who persuaded the Japanese military to use Stanley as the site for a civilian internment camp.

The camp had a resident Japanese commandant living in quarters not far from St. Stephen's College, but the daily running of the camp was left largely to the internees themselves, who formed committees to allocate the supplies provided by the Japanese and to handle internal administrative matters. Committees administered the canteen, looked after the education of children, regulated the use of electricity and looked after internee welfare and billeting, among other things. Presumably it was the welfare and billeting committees that ensured that I, a juvenile orphan with no mentorship or sponsorship, received support, protection and accommodation in a safe environment.

The three main committees reflected the main nationalities: American, British and Dutch. I came under the British Committee, which was later named the Council.

The Temporary Committee formed in January and was replaced by the British Communal Council in March. The Temporary Committee held their first meeting on 24 January 1942, which took place before my arrival. The British Communal Council elections were held on 8 February 1942, almost coinciding with the date of my arrival in camp.

§

The significant changes in my circumstances upon entering the internment camp were mostly for the better. Where I had endured loneliness, I now had all kinds of people to interact with daily; where there were space restrictions, now there was wide open space to roam freely around; because my life was unstructured, freedom was also undisciplined; above all, I felt secure! What more could a 10-year-old wish for? Did I profit from all these benefits? Probably not in the long term, but while I had them, I enjoyed them. On top of it all, here I had a place to sleep and food to eat, but clothing was a problem not only for me, but many others. In short, I felt happy and totally secure.

I have very little recollection about rations and food during the period that I had spent in camp. In the early months of internment, the quality of food supplied by the Japanese was considered acceptable. However, in the early days of internment, many people were still supplementing their camp rations with their own canned food and supplies they had brought into camp with them; obviously, I was not one of them!

Rations from the communal kitchen were served twice a day in the central courtyard of the Married Quarters at 11:30 a.m. and 5:30 p.m. The main staple was cooked rice, supplemented with a vegetable and maybe some fish or a meat. There was also a morning delivery of congee at around 8 a.m. Because of the possibility of cholera, dysentery and worms, all vegetables, including lettuce, were thoroughly cooked. I can still remember gagging while trying to swallow slimy cooked vegetable leaves; this was from a broad leaf vegetable with a furry underside. There was not much of meat or fish, and at times they weren't too palatable, or even edible.

Rice cooking was carried out in large vats or woks in the communal kitchens. After all rations had been served, the kitchen staff would often have to scrape burnt rice off the bottom of the large vats in the cleaning-up process. As children, we quickly learned to hang around the kitchen; with luck, some of us might receive a handful of burnt rice to chew on. This was a nice supplement to help fill any empty voids in the stomach, but the crusts were usually hard on the teeth. Another unpleasant experience during the avid consumption of cooked rice was unexpectedly crunching down on grains of sand. It seems the Chinese contractor supplying the rice to the camp cheated the weight by adding sand into the gunny bags. Damaged teeth were quite prevalent in camp because of this; I know it wrecked mine.

From time to time, the Japanese authorities used to ship into camp recently-stripped cow carcasses with shreds of meat still clinging to the bones. The hospital staff used to boil the bones to prepare soup for patients in our hospital, and the cleaned bones would be 'roasted' in an incinerator, and afterwards pulverized into powder form. I well remember gagging while trying to swallow two tablespoons of this dry 'calcium' powder, washed down with cold water.

Mr. Cunningham was a kindly older man in Room 17, who was a pipe smoker. He took great pains to warn me of the perils of smoking. He showed me the dark colour of his wooden pipe and said that my lungs would take on the same colour, should I ever fancy indulging in this habit. I couldn't help but wonder what was compelling him to blacken his lungs, but thought better of questioning him because he was an adult. A common supplemental beverage at the time was tea made from pine needles. When they were spent and dried, Mr. Cunningham used them as tobacco in his pipe.

Also in Room 17, appeared to be a single lady whose hair was auburn in colour. She was probably in her late 30s or early 40s, and who reminded me somewhat of Miss Gibbins of the DGS. One day, I happened to be in the room when she opened one of her suitcases, and I was intrigued to see that it was full of bundles of banknotes. I am still not sure of her name.

I was totally surprised when, one day, completely out of the blue, Christine Corra suddenly showed up in my quarters, armed with a drawing pad and set of coloured crayons. Christine was very special to me, as I had met her one or two years earlier in peace time at a Sunday brunch party given by Mrs. Rodgers in their house on Coombe Road on the Peak. I loved Christine's warm and bubbly personality, and especially the interest she took in me. How did she know that I was in Stanley Camp? By one of the following two possibilities: seeing me in the 'chow' line, or she was living on the same floor in Block 5. Whatever it was, it was a welcome flashback to a happier time.

What made this meeting with Christine so special? She asked me to sit for a portrait session and drew me with coloured pencils. Miraculously, I still retain this treasured 10 x 14 sketch to this day. Apart from this activity, I don't remember seeing much more of Christine afterwards, even though she and her mother might have been billeted in the same building that I was in.

Since the announcement of Mother's death, I had developed a penchant to suck my thumb when I felt bewildered or under undue stress. I had carried this habit with me into Stanley Camp, but this was soon to be terminated.

Sketch by Christine Corra made in Stanley Civilian Internment Camp in 1942 and the note from the back.

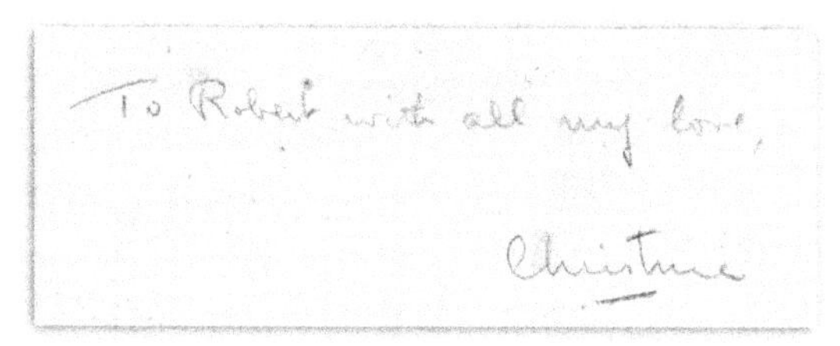

To Robert with all my love,

Christine

In the early weeks of internment, I occasionally came across Mrs. Grant sitting on a parapet near Block 2 or 3 by the Married Quarters. She was still quite portly at the time, but she always gave me a smile as I walked past her—maybe we even exchanged a few pleasantries. On this occasion, I must have been engrossed in deep thought as I was walking along. I suddenly heard a voice saying, "What is a big boy doing sucking his thumb?"

I spun around, looking for the source of this admonition, and there was Mrs. Grant, with a twinkle in her eye, giving me a broad but kindly smile. At that very moment I withdrew my thumb, and from then onward, I was totally cured of this habit! God bless you, Mrs. Grant!

Mrs. Grant was interned in Stanley Camp together with her three daughters, Rosaleen, Kathleen and Eileen. I subsequently learned that her three daughters had been pre-war students at the Canossian Convent, also known as the Italian Convent Sacred Heart School.

From the balcony of Room 17 (the middle floor) in Block 5, there was a commanding view over the high walls of the entire maximum-security prison below. One day, I was on my balcony taking in this view when I happened to glance upwards to the balcony above me (the top floor) and made visual eye contact with a lady looking down on me. I was totally startled, as the lady reminded me vividly of my godmother; I was almost certain that the lady was Olga Robinson! Due to an unexplainable impulse, I broke eye contact without exchanging any words and quickly withdrew into my room, wondering what would happen next. I couldn't explain why, but I remember being troubled by this 'apparition'. Nothing happened afterwards. I wondered whether it was all in my imagination.

After the war, on our way to the UK, I learned without a doubt that my godmother had been interned in Stanley Camp, and the lady that made eye contact with me had to have been her! Why was there no follow-up from her? Why didn't I talk about this to my godmother in subsequent years? I have no answers. Researching this further, many years later, I found that Olga Robinson, her mother Catherine Dobrjitsky and Harry Blake had all been forcibly detained with other prisoners by the Japanese in the unsavoury Toi Koon Hotel prior to their move to Stanley Camp in January of 1942.

If anybody should have been entrusted with my care in camp, it would have been Olga Robinson! Why that didn't happen at the time of my greatest need remains unknown to me to this day. This omission would be somewhat vindicated in the next phase of my life when we were united with each other after the end of the war.

§

My life in Stanley Civilian Internment Camp was idyllic. My daily existence reads like a story out of The *Boy's Own Paper*, a new adventure unfolding every day. I went wherever I felt like going, whenever I felt like going. I was subject to no adult supervision or discipline. The classes the education committee made available were rather limited at the time, so even attending school was optional! I had no curfew, and never had to stop what I was doing to do the bidding of another. I deeply commiserated with my friends who were subject to the misfortunes of chores or family mealtimes or set bedtimes. My new life was as close to heaven as a boy could wish. The negative impact of this kind of freedom would become manifest later in life.

In his book *Life in the Japanese Civilian Camp at Stanley*, Geoff Emerson wrote:

> *"With more than 200 children in the Camp as well as teachers and administrators from the Education Department, University of Hong Kong and a number of primary, middle and other schools, it is not surprising that very early in internment plans were made for education.*
>
> *At first, however, there was little enthusiasm from the internees for setting up a school. People were occupied with the more basic problems of food and housing, and also, there was a general feeling of resentment in the Camp against many of the parents with children. These parents were blamed for having disobeyed government orders in keeping their children in the colony. In addition, some internees felt that if the children were not in the Camp there would be more food and space for the adults. And, of course, almost no one expected to remain in the Camp long enough to justify establishing a school; people expected either to be repatriated quickly or to have the war end soon.*
>
> *Nevertheless, the educators were committed in their purpose and before the end of February 1942, kindergarten and transition classes began in St. Stephen's College Hall for children five to eight years old, junior school classes for eight to twelve years old and senior school for those up to eighteen. For the first few weeks the American children received separate instructions, but beginning in April 1942, the schools were combined."*

By the end of February, I was attending classes in St. Stephen's College Hall. The hall was used by both the junior and senior students. The path up the hill to the college that we all walked was opposite the Dutch Quarters, not too far from the guarded main gate into the camp.

I was assigned to the junior school, where classes were held in the mornings. We used to gather around our teacher as we sat in a small group on the floor of the main hall; and that was how our lessons were delivered. I can still remember busily erasing pencil writings on the back of salvaged labels from food tins, to be re-used due to the shortage of paper.

It was decreed that schooling hours for the elementary and junior students be restricted to two or three hours per day. There were two reasons behind this: firstly, there was not enough room to hold all the classes at the same time in St. Stephen's Hall; secondly, to conserve energy, due to inadequate rations for growing children. The second reason didn't hold up well, because children will be children, and they still burn up energy doing other things outside of classes.

I do not recall ever recognizing Miss Gibbins from DGS at St. Stephen's College during the entire period of my incarceration, but this may not be surprising, as the senior educational program wasn't in full swing until after I had left Stanley Camp. I understand that she was one of the professionals who contributed to this activity in a meaningful way.

Adventures with Wallen Winkelman

My closest friend in camp was Wallen Winkelman. He was in my class at St. Stephen's College Hall, was a year younger than me, and he and his older sister, June, were billeted in the Dutch Quarters with their American mother.

Wallen and I roamed around together when we had nothing else to do, which was often the case. One favourite spot was the cemetery, where we would lie on our backs on ancient granite tombstones, watching puffs of white clouds drift lazily across the blue sky as the wind sighed through the pine needles of the surrounding evergreen trees. The peace and tranquility were very soothing.

Despite Mr. Cunningham's warning, Wallen and I smoked when we could salvage rare cigarette butts on the pavement with still a few puffs left in them. These lucky finds didn't last very long, so there was never really any fear that our lungs would blacken from cigarette smoking.

Wallen was Eurasian and sometimes was the target of cruelty from other expatriate boys. I was with him one time when a group of boys started to taunt us. A shoving match ensued, and Wallen and I were crowded into a small concrete dugout, where these boys started hurling rocks and chunks of concrete at us. Despite successfully evading most of these missiles, eventually a rock struck Wallen on the head, and he started to bleed. The gang of boys immediately disappeared, which allowed us to clamber out of the dugout and seek medical attention for Wallen.

One breezy, hot afternoon, with Tytam Bay shimmering under a brilliant sun in a clear, blue sky, Wallen and I were walking along the footpath past the hospital when we met a Japanese officer and two armed troopers coming the other way. When the patrol halted in front of us, so did we. Then ensued an animated conversation between the Japanese as they looked at us. The officer detached himself from his patrol and strode purposely towards us, holding the scabbard of his sword in his hands. We were both scared stiff; Wallen took off like a rabbit, but I was frozen in place. The officer swiped at my left bare leg with the scabbard, almost knocking me over, and stomped on something next to my foot; it was a centipede that I had not noticed crawling up my leg. The patrol laughed at my fright and continued their way. Wallen came back, and I teased him about deserting a friend in need. We both had a good laugh.

We often strolled past the main gate of the maximum-security prison and used to wave at the soldiers inside the guardhouse. In time, we became acquainted with the duty officer in charge; he told us his name was Otaki-San. At times, he would invite us into the guardhouse and there he gave us some candy. I used to take some of this candy to share with my adult roommates, but most of them refused to accept anything originating from the enemy. Wallen and I had no such qualms.

On one of our visits, a staff car came through the main gate of the camp toward the walled maximum-security prison in the middle of the camp. Otaki-San shooed us under his desk, and we crouched there until the staff car had safely gone through into the prison grounds. Fraternizing with the enemy, even if they were only children, would have earned him at least a reprimand, possibly worse. I don't think

Otaki-San was his real name, because attempts to reconnect with him at the end of the war were unsuccessful.

Apart from Christine Corra, the only person I recall demonstrating a warm interest in me was Sister St. Stanislas de Kostka, one of nine Canadian nuns from the order of the Missionary Sisters of the Immaculate Conception (MIC) from Quebec. I think they were billeted in Block 4, facing the football pitch.

Tribute to Sister St. Stanislas

This religious lady and I very likely often passed each other in my building corridor, and even probably in the queue for our daily rations. This was how she took a personal interest in me, being alone. In time, I must have responded to her that I had had my First Communion some years ago. With that knowledge, she immediately started to prepare me for my Second Communion, which took place at Easter on April 5, 1942.

The Missionary Sisters of the Immaculate Conception are members of a religious congregation of women dedicated to serving the nations of the world most in need. Founded in 1902 by Délia Tétreault (1865-1941) in Canada, they were the first such institute established in North America. Members of the congregation use the post-nominal initials of MIC.

There were nine MIC sisters in Hong Kong when the Japanese attacked the city on December 8, 1941, and two of them were registered with the Auxiliary Nursing Service. All nine were interned in Stanley Camp on January 20, 1942: Sr. St. Philippe (Annette Beaudoin); Sr. Marie de Georges (Corrine Crevier); Sr. St. Antoine de Padoue (M. Angèle Forest); Sr. St. Etienne (Aurore Plouffe) (ANS); Sr. Marie des Victoires (Joséphine Bolduc); Sr. Marie-du-St. Sacrement (Anna Bourbeau); Sr. Therese de l'Infant Jesus (Yvonne Gérin) (ANS); Sr. St. Stanislas de Kostka (Germaine Gonthier); and Sr. Jean-l'Eucharistie (Jeanne Moquin).

Easter Mass on April 5, 1942 was held in the prison officers' club building and included First Holy Communion. For me, it was my Second Holy Communion. Sister St. Stanislas had me cleaned up and properly dressed. I felt rather conspicuous being so clean and well-dressed for this special occasion. With no memory of my First Communion, I was awed by the solemnity of the proceedings. Afterwards, there was a party, with special treats for all the communicants.

Prayer book by Sister St. Stanislas for Second Communion in Camp. 1942

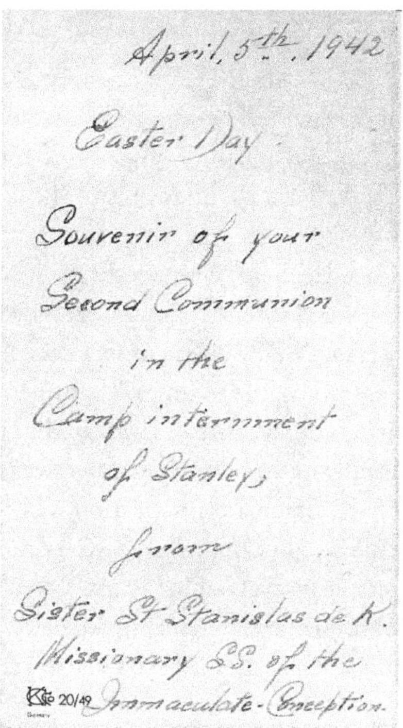

April, 5th, 1942

Easter Day

Souvenir of your
Second Communion
in the
Camp internment
of Stanley,
from
Sister St Stanislas de K.
Missionary SS. of the
20/49 Immaculate-Conception.

Sr. St. Stanislas' main motive in undertaking my religious education might have been to ensure that my soul would be properly taken care of in times of extreme peril. She kindly gave me several gifts to mark this special occasion, consisting of a handmade 68-page prayer book with handcrafted illustrations, a holy picture inscribed (in French) on the back by another nun and a holy picture of the Blessed Virgin with a signed message of remembrance. I have all these gifts to this day!

I had been exposed to various religions from a young age: Russian Orthodoxy, Roman Catholicism, Protestantism (Anglican) and Zen Buddhism. None of them meant very much to me in those early years, because there was no practice or teaching in my daily life to reinforce any one religion. The preparation for my Second Communion in Stanley Camp, directed by Sr. St. Stanislas, was impressive.

Sister St. Stanislas celebrated her Silver Jubilee of Profession in September 1942 in Stanley Camp at a ceremony officiated by Bishop O'Gara, specially to mark the event. She died at the age of 50 on September 3, 1945, having lived 30 years in religious life, and was buried in Canton at the White Chapel Cemetery.

My Exodus from Stanley Camp

As far as I can remember, no one expressed their concerns to me about my well-being when I was in Stanley Camp. In other words, I was left pretty much on my own to look after myself without any survivor skills. Perhaps, unawares to me, some of the adults in my room might have been keeping an eye on me. Was I told who to contact if I felt I needed help? Somebody must have initially guided me in collecting daily rations! This and much more I had to learn by trial and error, and without being aware of it, I could have been short-changed in the process from time to time. A baby gets attention with noise; I never wanted attention because of the fear that it might turn out to be restrictive—a totally natural but immature attitude on my part!

Camp administration was conscientious in meeting the needs of children in camp. There were other orphaned children there, and kind people who cared for them. What set me apart at the time was the lack of information about my identity. Being an orphan at 10 years old and alone in camp must have caused concern for the camp administration.

My unusual circumstances may have prompted the authorities—perhaps the British Communal Council or the welfare committee—to look closer at my situation to find a way to provide the support I needed. Whether their investigations were general or direct, they eventually found out that I had two sisters in the care of nuns at the Canossian Convent school, an institution well known to government circles.

After about six months of near-total freedom and independence, and without consultation or discussion with me, the authorities at Stanley Camp transferred the responsibility for my welfare directly to the nuns operating the Canossian Refugee Centre. This was very likely for my good, but from a 10-year-old's perspective, I didn't perceive it that way. To me, the British Communal Council in camp had traitorously negotiated with the enemy to dispose of one of their own British citizens, an orphan, and delivered him to an ally of the Japanese, the Italians.

I didn't want to go, and I pleaded to stay, but to no avail. Somebody in charge must have tried to assure me that this was for my benefit, and after all, I was to be reunited with my two sisters. I wasn't impressed at the time.

I had no recollection of seeing my sisters even once, since before Mother's death in 1939. Because of this, they were almost as good as strangers to me, and the idea of seeing them again held no allure, compared to the lack of direct supervision I was presently enjoying in Stanley Camp. Also, I didn't want to go to another boarding school, remembering the negative aspects of my experience at DGS for the last three years.

The thoughts going through my mind at the time can only be considered irrational. If only somebody had cared enough to gently win me over to this decision. I felt betrayed by all the adults around me. Had I not shown that I could care for myself, and perhaps with a little help from time to time, I wouldn't be a bother to anybody?

My protests went unheeded, of course. The Japanese cared little about my identity and later issued me an ID document that clearly stated in red print that I had 'enemy' status. I imagined that this was meant to restrict my mobility once I got out of Stanley Camp. My brief heavenly sojourn thus far came to an unceremonious end, and in mid-June of 1942, I was transported to the Italian Convent on Caine Road. All I could think of was leaving my newfound friends and the environment that I had grown fond of.

In the three years since my mother's death, I had been very careful to isolate myself from all disappointments in life, and perhaps this impeded my ability to speak up for myself. The principle that children should be seen and not heard placed me at a great disadvantage without a spokesperson who I could trust to speak on my behalf.

Heartbroken and angry, I huddled away from my adult escorts in the back of the car as we left Stanley Camp, silent and refusing to look at them. Bishop O'Gara might have been one of my escorts, and Dr. Selwyn Selwyn-Clarke the other, together with a uniformed Japanese driver. I hated them all.

Life in the Canossian Refugee Centre (1942 – 1945)

The convent site filled a stretch of land between Robinson Road and Caine Road on Mid-Levels in Hong Kong. The main entrance, on Caine Road in the centre of the complex, looked far more foreboding than when I visited the convent with my mother in 1936.

I carried my meagre belongings from the car and silently followed my escorts into the austere parlour, where we were received by Mother Teresa Zambaldi, the Mother Superior, and her attendants. My companions introduced me to the nuns and then left with the Japanese driver. I was next handed over to the care of Mother Rosetta (Lassini), who oversaw the child boarders. She then took me to my quarters.

The pre-adolescent children in my group who I remember were: my sister Julie, Therese Simmons, Theresa Stewart, Marie Gonnelli, Diana Chan, and on a temporary basis, there were other young boys, and I do remember Berkley with an American mother and Chinese medical doctor father.

The school property used to provide separate facilities for a kindergarten, a primary school and a secondary school. During the Japanese occupation, the school became a refugee centre and the school areas were repurposed. The east wing contained the nuns' quarters, which were close to the infirmary. The south-central block, the former secondary school, now housed refugees, as did the west wing, containing the orphanage and the auditorium. I was billeted in the west wing, where I had a bed in a segregated dormitory and assigned a school-type desk to stow my personal belongings.

The south-central block had a commanding view north across the harbour and towards the hills of Kowloon. The central courtyard had a grotto in which was a statue of the Madonna, and a small fountain in the centre. Completing the scenery were some trees and other foliage and a beautiful chapel building on one side of the courtyard.

For quite sometime afterwards, I still pined for my life at Stanley Camp. I missed the occupants in Block 5, especially Wallen Winkelmann, as well as the serenity I was beginning to enjoy. Compared to the freedom and airiness of Stanley Camp, the aging, close-crowded buildings of the convent complex really felt like a prison. The only bright spots in my view were the beautiful chapel building and its interior, and the open central courtyard.

One circumstance that brought back a twinge from the past was that I was no longer known as Robert, just as I was getting used to it; it had reverted to Robbie, because that was what my sisters always called me, and everybody else followed suit. Another positive experience was that I learned a little more about our family background from Margaret, and this knowledge increased my self-confidence, although I still had unanswered questions in my mind. However, the ID document issued to me by the Japanese cleared my mind about my identity; even though it didn't contain a full family history, I was classified 'enemy'.

With time, it wasn't difficult to rationalize the positive aspects of my life in the Refugee Centre. This was a place familiar to my mother—not that this information had too great an impact on me. The nuns knew Mother and the rest of the family intimately, going back many years in the past. The nuns were caring and supportive as best as they could be under these difficult conditions. All that was left was for me to benefit from association with the other refugees living in the same centre, and this was going to be quite rewarding. This realization would not come to fruition until after the war.

The Japanese authorities regulated and restricted written communications to internees in Stanley Camp. They provided 9 cm by 14 cm postcards on which one wrote greetings on one side only, and the recipient's name and address on the other. Once I was settled at the convent, I decided (for reasons I can no longer recall) to write to my godmother, who by now I suspected was interned in Stanley Camp. I never did receive a response.

§

Even though regular English schooling was forbidden by the Japanese authorities, I did receive private lessons of age-appropriate mathematics. Mother Maria Riva tutored me in the courtyard on a regular basis, and was very patient, as I initially struggled with the mathematical fundamentals of proportions.

I took Chinese language classes for a time, and my only regret is that I did not pursue this opportunity more vigorously. The teacher gave us all Chinese names, and this was mine:

達 巨 文

The first character is the surname (pronounced 'Dart', phonetically closest to 'Tatz'). The last character means 'Great', as in Great Britain.

The middle character is 'Achievement'.

I also enjoyed working with traditional brush and black ink calligraphy, and soon learned how to keep the forearm of my writing arm out of the wet ink on my pages. I learned the Chinese and Japanese national anthems and joined in singing them as required at the beginning of classes. The Japanese anthem was very mournful, but the Chinese anthem was like a march, with energy! These activities were optional, and I dropped out of them after a time. I am sure I had my reasons at the time, but now think it a great pity to have given up such an opportunity.

§

Mother Wilhelmina was a music teacher as well as an artist, and I eventually took art lessons from her. Under her teaching, and without pressure, I discovered my own artistic skills. Art classes were a pleasurable diversion, and I enjoyed producing artwork that pleased me and earned the praise of my teacher. I drew many pictures of Mount Fuji Yama with its snow-capped peak, and sketched human body parts (hands, feet, torsos, noses, ears, etc.). I was never good at drawing animals—horses were particularly difficult—but once I did produce a creditable reclining Bengal tiger with its head raised. With so much time on my hands, I also started reading, and spent hours at a time reading any book I could lay my hands on.

Mother Rosetta saw to it that I received pocket money from time to time. I used it to pay for regular haircuts at a barber shop on Caine Road. I was always nervous to be outside on the street, fearing to be stopped by Japanese patrols, even for a short outing. I also loved exploring the gift shop in the convent and spent any leftover money—and any I could pester Mother Rosetta for—on religious goods there. I still possess a treasured Sunday Missal, Latin-English Edition, and a plaque with an inspirational picture of the Sacred Heart of Jesus.

She usually said the grace at all meals and doled out the rations to each one of the children in her care. Evening prayers, especially the recital of the Rosary, was also led by Mother Rosetta. From time to time, she also led us on excursions outside the Refugee Centre. She was like a real mother to all of us.

§

The refugees at the Canossian Convent were a mix of enemies of Japan, neutral nationalities and Axis allies. The Canossian order had Italian origins, and Italy was part of the Axis (with Germany and

Japan), so the nuns enjoyed a certain degree of immunity from the Japanese authorities until the Germans were driven out of Italy. All resident refugees, though, were required to carry identification issued by the Japanese gendarme authorities. My identification declared my 'enemy' status with relevant sections highlighted in red. It was difficult to imagine what threat a 10-year-old could represent, or what damage he could inflict upon the mighty Japanese war effort. It was no easier to imagine the threat most of the other refugees posed, either—they were artists and engineers, dressmakers and teachers, parents and children, widows and orphans.

My sister Margaret told me that my beloved Ah Kai had sought refuge in the convent during the Battle for Hong Kong but was turned away due to overcrowding at the Refugee Centre. I never did find out what became of Ah Kai.

Deprived of the companionship of Wallen Winkelman, I made a new friend in fellow refugee Johnny Chan, whose family lived one floor above my billet. Johnny's father was the artist G.T. Chan, who had enjoyed international acclaim before the war. His mother was Italian. He had an older sister called Elisa and a baby brother called Victor who was born only last November, just before the war started.

After supper, I would go out to the back steps and whistle. Johnny would come out, often with a piece of his mother's special bread to share with me. I don't think his mother was even aware of this, but the bread was delicious, and I was most appreciative.

In her husband's absence, Mrs. Chan did dressmaking and taught French and Italian lessons to support her family and pay for rent in the Refugee Centre.

Johnny and I spent a lot of time together, wandering around the limited convent grounds, playing with other refugee children (boys and girls), getting into mischief on our own or huddling together underneath a makeshift canopy when it was raining, listening to the patter of rain drops overhead. Games of hide-and-seek were a favourite with all the children; games often only came to an end when the sun went down, or when someone carelessly knocked some fruit off a tree in the courtyard, leaving a mess we had to explain later to the nun in charge.

We invented our own game of submarines. This involved plastering mud around a wooden stick to weigh it down in water (and also putting the weighted sticks into a large, deep tub used to store rainwater). We'd watch them sink like submarines and then wait for the mud to dissolve and release the stick. The last stick to resurface won the battle.

Johnny and I sometimes dropped in on the Atroshenko boys, who lived on the same floor as Johnny. Viacheslav was our age and a gifted pianist who gave concert performances in the school auditorium. His brother Paul, about five years younger, was rambunctious and funny. Their parents had fled Russia because of the 1917 revolution. Mr. Atroshenko was a mining engineer; Mrs. Atroshenko was a dressmaker.

At one Christmas time, a neighbour, Mr. Wilkinson, gave Johnny and me each a new toy revolver. We had no idea where he got them from, but we were delighted and grateful. They were the most magnificent toys either one of us had ever received—authentic-looking and beautifully made, with a six-shooter barrel that took round caps, containing six shots each. We spent long, happy days swaggering around with the guns thrust into our waistbands, pretending to be Hoppalong Cassidy or the Lone Ranger. Imagination supplied the sound of horses neighing and galloping around the convent property and into the imagined sunset.

Johnny—who later became known as Tony—and I are lifelong friends. Despite a lapse of contact for over 67 years, we rekindled our friendship at his home on Vancouver Island and enjoyed retelling anecdotes of our time together in the Italian Convent.

Claude, Paula and Sydney Hollands were billeted on the top floor of the central block, which had a sweeping view of the entire harbour. My sister Margaret was also billeted in the same space. Their brother, Buster Hollands of the Hong Kong Volunteer Defence Corp (HKVDC), was in the military internment camp. I visited them often in the convent and played board games. Claude used to make me laugh by making funny faces. Sydney was more serious but still good fun; he used to invent games to play with me. I played a lot of chess with them, and in due course became quite good at it. I have no recollection how long they took refuge in the convent before they moved out to join their mother sometime before the end of the war; she was in another centre somewhere else in the city.

Mrs. Tyrtoff was a Russian lady who lived in a room on the floor above mine. She was a widow, and her late husband, a high-ranking Russian naval officer, had served under the Tsar. I visited her frequently to hear her talk about her husband's adventures in the Russian fleet based in Vladivostok during the height of Tsarist influence. The Japanese fleet destroyed two-thirds of the Russian navy at the Battle of Tsushima Strait in 1905. Mrs. Tyrtoff's husband might have died in this battle, the greatest and most important naval event since Trafalgar. I believe her daughter was married to one of the managers of the HKSBC, and they might have been interned in Stanley.

Mrs. Costello, a prim elderly lady, always well-dressed but with a rather serious countenance, had a room on the same floor as Mrs. Tyrtoff. She was a kindly and friendly person and I enjoyed visiting and chatting with her. She sometimes admonished me not to go around looking serious and frowning; she showed me the creases on her face to indicate the potential consequences.

§

Freedom of movement on the streets of Hong Kong was very restricted, so outside excursions for pleasure were not frequent. The nuns did take a group of refugee children out from time to time, and because the nuns were known by the Japanese as belonging to an Italian order (Italy being an ally of Japan at the time), Japanese patrols did not bother the nuns or the refugee children in their charge.

Honeyville Outing

One summer, the nuns organized an outing to Honeyville, their house in Pok Fu Lam, close to the sea. Honeyville had been donated to the Canossian Mission by a Portuguese philanthropist, and the nuns used it as a retreat and as accommodation for blind girls in their care.

The group of nuns and children, including Johnny and I, walked from the convent along Caine Road, past Bonham Road and Queen Mary Hospital, for about an hour. Our destination in the shadows of Mount Davis was a beautiful spot, and what a change it was from the confines of the convent! We stayed there for about a week.

There was no beach, only a rocky shoreline, and that is where I learned to swim. My lessons were taught in the 'sink-or-swim' style. I had been challenged to jump across a stretch of water, from one rock to another whose flat top lay submerged under a couple of feet of water. I gauged the distance, assessed the effort required and leaped bravely with eyes tightly shut. When my feet landed on the submerged rock, I opened my eyes and stood up in waist deep water. I was pleased with myself, but now came the test to get back to where I had started. With considerable coaxing and encouragement from the other children, and with no other option, I took a deep breath, closed my eyes again, plunged forward off the submerged rock, took two or three thrashing strokes and climbed out safely onto the target rock. It was not an elegant swim, but that didn't matter to me: I made it!

One morning, six or seven of us went for a hike in the hills across the road from the house. The well-beaten path up the hill meandered through an area where stands of bamboo had been cut down with sharp machetes, leaving spear-like stumps protruding two or three inches from the ground. I stepped on one and it speared through my tennis shoe and into my instep. I limped along painfully but soon dropped behind the others. Our route took us past a hornet's nest on the ground just off the path. The vibrations caused by six pairs of feet close together agitated the hornets, and they eventually directed their displeasure at me, delivering a couple of painful stings as I limped into their midst. I streaked down the hill, overtaking everybody without any problem, in spite of my wounded foot. I didn't stop until I reached the relative safety of the road. Some situations can be a great incentive for action!

Another day, Johnny and I went exploring in another section of the nearby hills, and we stumbled across bleached skeletons of several human remains. There was some evidence they might have been looters, because remnant clothing did not appear to be military garb. We tucked a skull into a rattan basket, with the idea of scaring some of the girls back at Honeyville. The adults in charge found out, and of course, we were severely reprimanded for what we did and had to replace the skull where we found it.

Tiger Balm Gardens was like a fairyland to my eyes, with its many beautiful statues and the scenic view of the place from the top of the pagoda. Even the pool had statues—mermaids lying on the bottom—though there was no water in the pool.

Our excursions to the Gardens may have been due to the largess of Aw Boon Haw, an industrialist who owned the Tiger Balm (Chinese liniment ointment) company. He was a significant benefactor to the convent during the war. In addition to his significant financial contributions, he often provided large quantities of rice as well as bolts of fabric to make clothing for the orphans. His tremendous generosity often came in the nick of time, as shortages loomed, or privations were being felt.

Another place we went on outings not far from the convent was the Botanical Gardens. In its wide-open spaces, we ran to our hearts' content. I don't recall seeing the small zoo, but the grounds were well kept, and even I, with no knowledge or interest, was impressed by the arrays of flowers and exotic plants. Close by was the residence of the former British governor of Hong Kong, now being occupied by the Japanese governor, under full guard by military sentries.

§

I remember receiving a Red Cross parcel, about 15" x 36" wrapped in gunny sacking. It was in 1943, and the only one I ever received throughout the war. I don't remember what was in the package, but these parcels typically contained an assortment of goodies, such as chocolate bars, powdered milk, sugar, peanut butter, coffee, cigarettes, an assortment of candy, hard cookies and numerous non-perishable items. I was happy to share this bounty with some of my companions, and what I couldn't use, I handed over to Mother Rosetta to share as she saw fit.

Catholic Religious Influence Increases

None of the religions I had encountered before the war—Russian Orthodoxy, Buddhism, Roman Catholicism, Anglicanism—had influenced me greatly. My second Holy Communion in Stanley Camp had made a pleasant, if temporary, impression. But at the Italian Convent Refugee Centre, I was immersed in the Catholic faith for the next three years, and that had a profound effect on me.

At the beginning, the practice of faith was foreign, but it soon became a comfortable routine. At the time, I didn't realize the benefits of this experience; they would only be understood later in my life.

I attended Mass at the Convent Chapel (service conducted in Latin) every Sunday and confession every Thursday. I joined in saying grace before all meals and participated in prayers according to the religious calendar. I received intense religious exhortation from various nuns and before long became an altar boy, serving Mass in the chapel and even occasionally assisting Bishop Valtorta at Mass in the Catholic Cathedral a short distance along Caine Road. It is a wonder I was not seriously inclined towards a religious vocation by the time the war ended three years later!

The priest I assisted at most of the Masses in the Chapel was Father Grampa, an Italian priest. He was about 50 years old but seemed ancient in my eyes; he must have walked to the convent from his quarters near the cathedral. We often had guest priests, including the bishop, to celebrate Mass at the convent chapel.

As an altar boy, I assisted the priest in donning his ceremonial vestments and attended to him at the altar, including lighting the candles at the main altar. On special occasions and at high Mass, the six main candles at the back of the altar were lit from a wick at the end of an extended pole. At the end of the service, the candles were snuffed out with a hood on the same pole. The senior altar boys always enjoyed the privilege of lighting and snuffing out the candles at the end of a service, which was a tricky and a fun task.

Only the priest could dispense the sacred host in those days. The altar boy held a gold-plated dish under the chin of the people taking communion, to catch any accidental crumbs from the wafers the priest placed on their tongue. It being wartime and wine not available, communicants made do without it.

In the early years, Christmas Midnight Mass was always celebrated three times in a row, one after another. At my first Christmas as an altar boy (probably Christmas of 1942), I accidentally took communion more than once, and thought I had committed a mortal sin. The priest who took my confession afterwards assured me that I would not be condemned to eternal damnation and gave me his blessing.

The crowning experience for any altar boy, apart from serving at Mass on Christmas and Easter, was to assist at the ordination of new priests; it is usually a lengthy affair, which includes Mass. On November 6, 1942, in the convent chapel, I assisted Bishop Valtorta at this very solemn event. The chapel was splendidly decorated for this occasion, which began at 7:30 a.m. The candidates, all members of the Dominican Order, received the tonsure (cutting of the hair to form a bald spot in the crown), two received the Minor Orders (acolyte, exorcist, lector and porter), three the deaconate or Major Orders (clerics preparing for priesthood) and two the priesthood. After such a lengthy ceremony, everybody was famished for breakfast, not the least were the altar boys.

§

Many of the priests at the cathedral were avid postage stamp collectors, and I soon developed the same interest, having caught the interest from them. They showed me how to soak glued postage stamps from envelopes, how to mount stamps into albums and the difference between loose sheets and albums. I received an outdated Stanley Gibbon catalogue to help me identify various stamps and learned about 'original flaws' that enhanced the value of some stamps. I learned a lot about geography by collecting and exchanging foreign stamps, and this inspired my desire to want to travel the world. Over 70 years later, I still have part of this collection.

Allied Air Raids

The Allies, especially the Americans and Canadians, were fully engaged in warfare in the Pacific at this time (the British were busy in India and Burma), and life at the convent was often disrupted by air raids and bombing. This often resulted in interruptions in the electrical and water supply, shortages of food, fear, hunger, destruction and death. This was real war, and real people died with every bomb dropped and every bullet fired. The air raids also brought out the looters, who used the uproar and chaos to distract from and conceal their thievery.

In the Refugee Centre, there was more than one designated area that people could take shelter during an air raid. There were many opportunities to witness air raids taking place, the main danger was obviously from direct bomb strikes, but also from flying shrapnel and stray bullets. A popular lookout for me was from the top balcony of the Central Building which faces north, giving a panoramic view of the harbour and wide stretches of Kowloon.

Watching the action taking place in the harbour below was always fascinating and frightening. I don't remember seeing many Japanese airplanes, but certainly the Americans came out in strength. It was in 1943 that the air raids by the Americans started to increase in intensity, and through the next two years, until the end of the war.

Also, watching fighter bombers streaking over Hong Kong's harbour in attack mode was a sight to behold; silver bodies twisting and turning to evade gunfire directed at them, then re-grouping to turn back for an additional run. We always could count on an air raid when Japanese shipping was in port. The American strategy was to disrupt all shipping supplies.

Airplanes soared and dived, skimming over the water from west to east to shoot at ships in the harbour, or coming in low over treetops and buildings, dropping whistling, fiery death as they passed. Some planes were shot down—a sad and gruesome fact of war. Pilots were very brave, indeed, no matter which side they fought on. Most of the action took place some distance from Central and the convent.

A more ominous strategy was when the big bombers came over to destroy land installations, including shipyards, factories, power plants and anything useful to the Japanese war efforts. 'Carpet bombing' created huge destruction and severely impacted civilians.

The harbour held many partially or fully sunken ships from other raids. Some of these ships capsized, exposing the bottom of their hulls to the sky. It was on these hulls that the Japanese mounted anti-aircraft guns to shoot at the bombers.

In July 1943, 12 American bombers raided Taikoo Dockyard and parts of the city. More raids followed in August. In one of these raids, a bomb hit a building on Ice House Street, not far from the Refugee Centre. The explosion created quite a shock wave in the area, which was felt in the convent. After the all-clear siren sounded, Johnny and I snuck down to Ice House Street to see the damage. The building was a terrible mess, and human remains hung from the branches of some of the nearby trees. I saw a leg hanging on a branch. We didn't hang around long, because I didn't want the risk of being accosted by a Japanese patrol.

We learned later that a little girl called Elsie, who was one of the refugees at the convent, lost her mother in this bombing raid. I hoped she was not one of the people whose body parts had to be picked out of the trees.

That November, the water supply was cut off to the Convent district, perhaps by a damaged distribution system, and we were forced to rely on the water well in the convent grounds for a short time.

In December came another heavy bombardment, resulting in huge damage in Kowloon. Casualties were heavy, including the death of one Canossian sister. In October 1944, 30 American bombers raided the 20 Japanese ships in port, including Hung Hom. There were about 7,000 casualties.

One evening, just before Christmas 1944, I was watching a Japanese destroyer anchored in the harbour shoot down the middle aircraft of a trio formation flying in at low altitude from the west. The two outer ones peeled off immediately and flew back in the direction they were coming from. The raid had been so brief that the air raid siren hadn't even been activated.

In January 1945, a heavy raid with 30 bombers battered Wanchai, destroying about 500 houses; casualties were tallied between 15,000 and 20,000 people. The Americans also bombed Stanley Camp by accident; 14 prisoners were killed in Bungalow "C". After the war I learned that Uncle Harry, who was on his way to seek shelter in a stairwell during the raid, had his right ankle smashed by a ricocheted heavy calibre machine gun bullet fired by one of the aircraft. He was apparently bleeding profusely and nearly lost his foot. At the end of the war Uncle Harry was transported by hospital ship to England for proper medical attention.

In April, another massive air raid on ships in the harbour and land targets destroyed the French consulate building, among others.

When the war in Europe ended on May 8, 1945, the Allies turned their full attention to ending the war in the Pacific. American bombers dropped incendiary bombs on Hong Kong on June 12, 1945, resulting in great misery for the Chinese population.

Almost everyone in the city was hungry, if not starving. The Japanese currency had been revalued, and the price of available food supplies skyrocketed. Misery was over all Hong Kong, and dead bodies due to starvation had to be frequently removed from the streets. Things looked dark indeed.

We daily witnessed Chinese men, women and children lying on the pavement or huddled in doorways, dying gradually of starvation; their eyes, from which all expression had already left, looked at passers-by, who were powerless to assist.

And then, on August 6, 1945, the United States dropped Little Boy, an atomic bomb, on the city of Hiroshima. Three days later, they dropped Fat Man on Nagasaki.

There was talk of a Japanese surrender. The idea that we would soon be rescued, that troops were finally not far away, still fighting on Okinawa in Japan, raised our spirits and buoyed our hopes. We had lived with hardship for the last three and a half years; could it be the beginning of the end of the war with Japan? We held our breath until the official announcement followed six days later.

Japan Surrenders

On August 15, 1945, Japan agreed to an unconditional surrender to the Allied forces surrounding the islands of Japan. Because of the suddenness of the Japanese capitulation, Hong Kong became a political 'hot potato'. General MacArthur was in favour of the Chinese Nationalist government taking possession of Hong Kong. Several years earlier, US President Franklin Roosevelt insisted that colonialism would have to end, and promised Soong Mei-ling (Chiang Kai Shek's wife) that Hong Kong would be restored to Chinese control. But, Great Britain was not in agreement. It became urgent for Great Britain to dispatch a naval force as soon as possible to take possession of Hong Kong. This happened after a delay of about two weeks.

With so much uncertainty and a perceived lack of law and order in the meantime, the looters became active again. This time, their main target was the Japanese. The Japanese military were still armed and could defend against looters, but isolated Japanese who met groups of Chinese on the streets were sometimes savagely beaten.

By August 28, the supplies the Japanese authorities were holding back were released, to the great relief of the starving population, including those living in the convent.

Now that War is Ended - What Now?

At 13 years old and under the circumstances, this question never occurred to me. All I could focus on was that there was something big happening outside in the world that could have a profound effect on my life; what it was going to be I could never imagine! Practical considerations never entered my mind!

The kindness of the relieving military personnel at the time of the liberation was going to overwhelm and fill me with happiness beyond my memory. I felt it was too good to be true and before long I would wake from this dream! Instead, it got even better!

This is me in the courtyard of the Canossian Refugee Centre on Caine Road. The photo estimated to be in 1943 or 1944. At the time of the liberation, it was from here that I was notified that I would be repatriated to the UK.

CHAPTER 9: LIBERTY AND REPATRIATION TO ENGLAND

Happy Days Again

August 1945 – December 1946

The British moved quickly to regain control of Hong Kong. As soon as he heard word of the Japanese surrender, Franklin Gimson, Hong Kong's colonial secretary, left Stanley Prison Camp and, in downtown Hong Kong, he was sworn in as "officer administering the government." A government office was set up at the former French Mission building in Victoria on 1 September 1945. British Rear Admiral Sir Cecil Halliday Jepson Harcourt sailed into Hong Kong on board the cruiser HMS *Swiftsure* to re-establish the British government's control over the colony. On 16 September 1945, he formally accepted the Japanese surrender from Maj.-Gen. Umekichi Okada and Vice Admiral Ruitaro Fujita at Government House.

I watched from my vantage point in the central block in the Refugee Centre as the British fleet steamed into Hong Kong's harbour on Thursday, August 30, 1945, after minesweepers cleared the approaches to Lye Mun Pass. Between the last few days of August and the first week of October, the following list are names of some of the warships I remember seeing anchored in the harbour:

HMS *Duke of York*, battleship; HMS *Anson*, battleship; HMS *Swiftsure*, cruiser; HMCS *Ontario*, cruiser; HMS *Indomitable*, aircraft carrier; HMS *Indefatigable*, aircraft carrier; HMS *Vengeance*, aircraft carrier; HMS *Venerable*, aircraft carrier; HMS *Ocean*, aircraft carrier; HMS *Maidstone*, submarine tender; HMS *Adamant*, submarine tender; six S-class submarines tied up alongside Tamar basin;

HM *Oxfordshire,* hospital ship; HMS *Helford,* frigate; HMCS *Prince Robert.*

With constant movement of naval vessels, it was sometimes difficult to keep track of the flow of traffic. For example, Admiral Bruce Fraser was passing through on his flagship, HMS *Duke of York,* on his way to Japan to join up with General MacArthur's forces, in preparation for the official surrender from Japan in Tokyo.

At the earliest opportunity, and most days, I left the Refugee Centre to go down to HMS *Tamar,* the naval shore base located in the Central District. Naval personnel welcomed me and quickly ushered me behind closed gates to keep out the rampaging Chinese mobs who were seeking stray Japanese personnel and looting whatever was unprotected. I had seen the mobs' harsh treatment of the Japanese on the streets and had also witnessed the rescue of Japanese personnel seeking safety from vengeful Chinese citizens.

There was continuous movement of Allied vessels entering and leaving the harbour. I went down quite early every day and sometimes returned at night to the Refugee Centre, happily exhausted. One morning, I encountered a small, sombre procession pushing a wheeled hospital stretcher along Caine Road, going west. A white sheet covered a corpse on the stretcher, but the face was left exposed. A sword in its scabbard lay beside the man. Four grim-faced, armed Japanese soldiers marched on either side of this stretcher. The occupant was a Japanese army officer who had committed hara-kiri. I felt sad and sorry for that soldier all day.

During one of my many visits to HMS *Tamar,* the naval shore installation in Central District, I was invited aboard one of the six S-class submarines berthed alongside each other, which were collectively supplying power to the naval dockyard facilities. I toured inside HMS *Swordfish,* intrigued by all the gleaming instruments and the cramped living quarters. Unfortunately, there was too much vibration from the diesel engines driving the generators to allow them to raise the periscope for me to look through.

Rev. Father Chatterton, a Royal Navy Catholic chaplain, took an interest in me. He introduced me to a few naval personnel, who in turn arranged an awe-inspiring ride by naval launch to various man-o-wars anchored in the harbour.

On board HMS *Helford,* a river-class frigate, I was received by Captain Cuthbertson and his Executive Officer, Lieutenant Eric

Collection of Olga Robinson, photographer unknown

Stanley Civilian Internment Camp at time of liberation at end of August 1945. This was where the flag raising ceremony was carried out. The nearest building in the background is known as Block 2.

Reference: DND/DHP. Photographer unknown, forprosperityssake.ca

HMCS *Ontario* in Hong Kong Harbour at liberation. c1945.

Smith, and Brian Barlow. I was introduced to all the officers and then taken on a tour of the vessel. My exposure to the other ranks was when I volunteered to participate with painting of the ship's guard railings. It was a great feeling to be appreciated and gave me a great sense of satisfaction. It was here that I first entertained the idea of making a career in the Royal Navy.

On another day, I hitched a ride on the back of a military dispatch motorcycle going to Stanley Camp, which was still in operation due to lack of accommodation in the city for the freed ex-internees. The new colonial administration had asked ex-internees to remain in camp until they were called up for repatriation to their country of origin. The sick and maimed were given top priority, and relatively able-bodied RAPWIs (Recovery of Allied Prisoner-of-War and Internees) would follow.

While they waited, those still in camp had full freedom of mobility and a bountiful supply of daily rations. We stopped by the canteen building and were quickly surrounded by a crowd of ex-prisoners. The driver gleefully handed out treats to the children, and I looked around for anyone I might know. But the children had grown and the adults were thinner and looked older than three years should account for. I returned to town in Central with the dispatch driver without recognizing anybody on that visit.

I was in 'seventh heaven' the whole of that September, spending time at the naval yard and receiving continuous kind attention from the naval personnel all around me. I seldom thought more than a day ahead. It did not occur to me to try to connect with my godmother, whom I had written to from the convent several times without a reply. I rarely, if ever, saw my sisters during this period, either. Margaret was 19 years old and Julie was 15. I think they were still in the convent and suspect that they too were enjoying their newfound freedom of movement and life outside the convent. I had no idea how they spent their time. I was a month away from my fourteenth birthday and was happily spending my days at the naval yard in the company of honourable men and as a guest on various warships.

Repatriation to England

One evening at the Refugee Centre, M. Rosetta told me she had received orders that I was to be repatriated to the United Kingdom, along with my sisters. I could scarcely believe it! Were we being sponsored by somebody from Stanley Camp? Was this my godmother's handiwork, or that of the welfare committee in camp? I had no idea then, nor have I to this day ever known, whose invisible hands worked on my behalf.

RMS *Highland Monarch* Departing Hong Kong

On the morning of Wednesday, October 5, 1945, Margaret, Julie and I bade farewell to the nuns who had cared for us the last three years and made our way to RMS *Highland Monarch*, the ship that would transport us to the UK. We joined a group assembling in downtown Hong Kong—men, women and children ravaged by three years in a prison camp, excited to return to their families in the UK and return to a normal life again. The mood was festive and busy. Women and children under the age of 13 were assigned a berth in the mid-ship cabin accommodations. All military personnel, civilian men and boys aged 13 and up were assigned to the 'tween decks—the space under the main deck, where I would end up.

I was too excited to feel any nostalgia or regret about leaving Hong Kong. I looked forward, not back, as the Highland Monarch pulled away from her berth and made her way through the harbour, heading for Lye Mun Pass. Once through the Pass, the ship picked up speed and headed southwest towards Singapore, where we would stop to pick up more passengers. It never occurred to me that I might never see the Nine Dragon Hills of Kowloon, the Peak and mountains of Hong Kong Island again, but I also had no idea what lay ahead for me in England.

The ship was originally built before the war for the beef trade in Argentina, and the 'tween deck cargo holds were mainly designed for the carriage of refrigerated cargo. For this purpose, overhead refrigeration piping had been installed. When the ship was requisitioned as a troopship in 1940, modifications were carried out to make the ship suitable to carry troops, and this piping in the 'tween decks was very useful for slinging hammocks. Officers were accommodated in the midship first class cabins, where women and children were also given accommodation.

Royal Mail Liner Highland Monarch as a troopship transporting RAPWIs to the UK for recuperation after the war in Hong Kong. I was a passenger and it was here that I discovered that my godmother was also a passenger together with her mother. Sometimes I wonder if she ever knew that I was on the passenger list before we met. c. 1945.

The Highland Monarch's 'tween decks were divided into separate living areas, allocated by the military commanding officer and his staff on board the ship, according to the origin of each group of men: #6 Commando Brigade, #1 Royal Marine Commando Brigade, HKVDC personnel, other military regiments and service corps, and civilian RAPWIs. I was thrilled to be in close company with soldiers and listened rapturously to their stories of adventure of fighting the Japanese in the jungles of Burma.

Weather permitting, most of the men spent their time out on the open deck. Civilian men connected with their wives and children during the day in their midship cabin accommodations. Entertainment was also provided on the open decks and in the ship's lounges, which were open to everybody. Space was, of course, often very crowded.

Each night at bedtime, I wasn't very happy slinging my hammock down in the 'tween decks below. This was a tricky operation for me, being small and not particularly strong. Ventilation wasn't the greatest either and added motion due to unsettled seas could bring on seasickness. Rather than struggle with those conditions, I acquired a paillasse and slept outside on deck when it was not raining.

In addition to the normal lifeboats provided, our ship was also equipped with a series of stacked wooden life rafts, not unlike small swimming rafts. The idea behind them was if the ship was sinking, these rafts would float off the deck and be readily available to survivors in the water. I found a place on top of one of these stacked life rafts, and this was where I laid my palliasse. An added feature that I found most useful was that the top raft was smaller than the one below it, and this provided a nice ledge that I could snuggle up into with my 'sleeping bag'. All life rafts were positioned alongside the ship, and I liked to face the ocean as the ship ploughed through the seas, enjoying the night breeze caressing my exposed face. If the rain was slight, I would snuggle deeper into my 'sleeping bag'. I was usually not disturbed when, every morning, the crew would be washing down the deck around my little 'island'. It was perfect!

Meal arrangements for the men were simple. Food was brought in from the kitchen in large containers and set on a long table. We helped ourselves in buffet style and sat anywhere at the mess tables. Empty plates and utensils were deposited in a conveniently located bin, from where they were removed by the kitchen staff. Some men disposed of their used plates through the open porthole and into the ocean outside! Very wasteful, I thought!

§

I was quite surprised to see Mrs. Kennedy-Skipton and her two daughters as passengers on the ship. In addition, there was the large family of Mr. Wilkinson, who I had met in the Canossian Refugee Centre.

Separated from my sisters aboard the ship, our usual lack of closeness continued. Our interest in each other was warm enough but had no foundation built on intimate affection. I have no memory of spending any time with them on the ship. Any loneliness I felt was easily countered by the attention of the soldiers I was billeted with. They cared for me, played games with me, took me with them on shore excursions and showed me great affection. Every one of them was a hero to me.

Many of the soldiers did leather crafting, and they taught it to me. Under the tutelage of one soldier, I crafted a pretty good wallet. The embossing of a design into the leather came out quite nicely, and I was proud to show it off. It was also a handy place to keep some of the money I had received on my birthday.

Having only a very cursory idea of the power of money, I was not very careful with the little cash I received as an allowance on board the ship. One day, I lost my new wallet, containing a 10-shilling note, and imagined it gone forever. At the urging of my friends, I went to see the Troop Commanding Officer in his office, and there I found my wallet, which had been turned in by somebody. Was I happy!

§

Surprises continued aboard the ship. I can't remember how I discovered that my godmother, Olga Robinson, was also a passenger on board the Highland Monarch, but the knowledge brought back some negative memories. I was still somewhat bothered by her treatment of me when Mother was in hospital and after Mother died, when I was living in DGS. Also, why didn't she reply to my notes I had sent to her in Stanley Camp? These were unresolved, disquieting thoughts I didn't want to dwell on in my present state of happiness.

The ship was steaming in the Bay of Bengal towards Ceylon, when imagine my surprise to receive a note from the Chief Steward, with a written invitation from my godmother to attend dinner in the ship's main dining room in celebration of my 14th birthday on October 11. I had forgotten all about my birthday on that day because I was beginning to feel motion sickness.

The weather had been changing during the day, and the ocean swells were getting larger. My queasy stomach worsened steadily, and by dinner time, I was decidedly unwell. Because there was a sudden lurch of the ship as I stepped on the stairs to the dining saloon, I stumbled a little and had to grasp the handrails firmly to steady myself. Then the white-uniformed steward led me to my chair, and at the table was my smiling godmother, together with some other passengers. In my developing sea-sickness, I was barely conscious of my godmother's greeting of welcome to the table as I was devoting all my waning energies to avoid looking at any portholes, because the moving horizon would greatly add to my misery.

I had tried to control my heaving stomach, but there was no arguing with seasickness. Before I could finish my first course on the menu, which was mock turtle soup, I was forced to dash out of the dining room, up the stairs and onto the moving deck. I clung to the guard railing, hung over it and violently fed second-hand mock turtle soup to the denizens of the deep.

I didn't return to the dining saloon to eat my unfinished birthday treat but wobbled back to the 'tween decks down below to nurse my affliction. My companions were amused by my condition and offered remedies and advice for my relief when all I wanted was to be left alone to die peacefully! I was perkier the next morning, and before long I was eating, drinking and taking part in activities as usual. From then onward, I carried on like a seasoned sailor and was never seasick again for the rest of the voyage.

§

Our stop in Colombo, Ceylon, was a major event. The Red Cross had set up a station on shore to outfit all the RAPWIs on the ship with summer clothing and shoes. For the first time I could remember, I got new underwear to go with cotton shirts and shorts. It was a new experience to be free to choose anything I wanted. It was stunning to be outfitted so lavishly. Talk of a kid turned loose in a 'candy shop'!

Gracie Fields, a singer and entertainer, came on board the ship to put on a show. The show was well attended by all the passengers who were crowding against the ship's railings and all available space. I had never heard of her before, but she was a favourite with all the soldiers. She told jokes, sang her heart out, got the audience participating and she made everybody happy.

The ship docked at Port Tewfik in Egypt, at the entrance to the Suez Canal, to furnish all the RAPWIs with a winter wardrobe. The Red Cross had set up a large Quonset hut in the port area, where they again helped us to select new clothing, this time against the winter in the UK. We were given flannel trousers, warm underwear, coats, gloves, hats, scarves, woollen socks and heavier shoes. Toiletry supplies and towels were also included. At this stage, I think I had need of a carry-on bag, large enough to handle all the stuff I had accumulated since Colombo. I was very happy, because life was exceedingly good to me!

Within the secured dock area, where clothing distribution was located, RAPWIs were only permitted to wander in the immediate area near the vicinity of the ship. Military personnel in uniform were also allowed to leave the dock area and visit the nearby city, which I think was named Suez. At their invitation, I decided to join my military friends on a planned trip into Suez, not realizing the restriction imposed on the civilian passengers. Of course, when we reached the exit check point, the military police there turned me around and sent me back to the ship. I was not too happy about this, but it was a lesson learned.

The next inbound vehicle with room for me was chauffeur-driven and contained an important-looking Egyptian wearing a suit and tie and a strange hat with a tassel—a fez. I shrank as far away from him as I could in the rear seat of the car, and he only laughed and spoke teasingly to me until he dropped me off at the ship's entry station at the dock.

§

I remember little about the passage through the Suez Canal, except that the scenery didn't seem very interesting, there was only desert to see on each side of us. We passed Port Said at the northern end of the Suez Canal at the end of the day and then entered the Mediterranean Sea, where the weather became decidedly cooler.

We stopped briefly in Gibraltar, I think for the ship to take on fuel, but long enough to allow a short excursion into the town. The visit into town was with my godmother, and all I can remember was that Gibraltar was a huge rock populated by wild monkeys.

Leaving Gibraltar would begin the last leg of our voyage, as the ship entered the North Atlantic Ocean, and soon it was on course north along the coast of Portugal. The Bay of Biscay ahead of us had a notorious reputation for rough seas in winter but imagine our surprise to find the seas as placid as a lily pond; this was to continue all the way to our destination.

As Southampton loomed ever nearer, the seagull activity following the ship increased in intensity. I was struck by the verdant greenery on the Isle of Wight, and soon the Port of Southampton appeared to our left. I gasped in awe to see the RMS *Queen Mary* docked close to our berth, still painted in its wartime colours. On the morning of Friday, November 9, 1945, I finally arrived in England! Then I had to hurriedly complete preparations to disembark.

With more than 800 passengers eager to disembark from the ship, there was a lot happening all at once. Without any leadership from my oldest sister, I unconsciously drifted towards my godmother, who was close by. I could see she needed help, as she had to take care of her aged mother as well, who had no English. She welcomed my offer of help and from then on, we remained close companions, which was to endure throughout our stay in England.

Arrival in England

The British Red Cross transported a group of us into London to our billet in the Notre Dame Rest Centre. This centre, just off Clapham Common, was operated by a Catholic religious group of nuns. The Red Cross representative, Marjory Kirby, assigned to look after us at the Centre, eventually became a very close friend of my godmother for many years. Marjory worked tirelessly to ensure that our needs were met and that we were comfortable. I think we stayed here for several weeks.

I remember meeting Catherine Hellevik for the first time in the Norwegian Club at the Shaftesbury Hotel in London. She was a good friend of my godmother from Stanley Camp. Catherine was in mourning, having recently lost her 14-year-old son Norman when they were on a repatriation ship from Hong Kong on the way to Australia. For many years afterwards, we used to correspond with Catherine, who was then living in Bergen.

Norman Hellevik and his mother were interned in Stanley Camp during the war. He was diabetic and under the direct care of his mother. The wartime diet in camp was no threat to Norman's diabetic condition but, separated from his mother on the repatriation ship, it appears that he tragically succumbed to a severe diabetic coma. His burial took place from the ship when it was at sea. It was such a tragedy to survive three and a half years of captivity and then, a month after liberation, die at such a young age!

One day, I took a train by myself to visit Derek Smith, the former executive officer on HMS *Helford*, at his home in Morden, just south of London. Derek was one of the officers who had shown me hospitality and kindness on his ship after the liberation of Hong Kong. I was introduced to his parents and sister Paula and had tea with the family at their home. Like an older brother, Derek gave me a lot of advice about life and had urged me to stay on course to complete my education, which had been so badly interrupted by the war. Of course, he wasn't privy to my circumstances (something I was reluctant to share with anybody) and he couldn't know how vulnerable I was at this point in my life.

I experienced another first in London. This was my visit to the fish and chip shop with my godmother in Clapham Common. The store was one of a few situated on one side of the Common Green. I think the fish we ordered was cod, but it could have been plaice. It was served piping hot, all wrapped up with potato chips in a newspaper, probably The News of the World. It tasted delicious, no doubt enhanced by the printing ink from the newspaper. In England, and especially London, practically every day would bring new experience my way!

After our time was up in London, Marjorie Kirby came around to assist us to make the final move to our destination in Coventry. It was at Euston Station that we boarded the LMS train to Coventry. Marjory and her husband Reginald became close friends, and when I was on my first long leave from Jardines many years later, I visited them in Saffron Walden, at which time they had a young daughter named Anne.

One disappointment in leaving London to go to Coventry was that Godmother would be further separated from Uncle Harry, who was lying in a hospital in Winchester having his right foot looked at by specialists. This was the foot that was hit in Stanley Camp by the Americans in one of their air raids on Hong Kong in January of 1945.

Baginton Fields Hostel – Coventry

In early 1945, the Baginton Fields Hostel in Coventry housed a group of Dutch children recuperating from the war in Europe; they might have been wartime orphans. These children had been re-settled with English families by the time the RAPWIs from the Far East became guests of the UK Ministry of Health.

A helpful woman in a station wagon met us at the Coventry railway station and drove us to the hostel, which was about four miles outside the city. The main building housed the administration offices, dining room, kitchen, lounge and auditorium for gatherings and dances on the weekends; there was a separate laundry building. Clustered around these facilities were about 10 accommodation units, or blocks, alphabetically designated. Each block had about 10 bedrooms (each bedroom had two or three beds), each equipped with a sink. Each block included a common ablution room with toilets and showers. Paved walkways connected all the buildings. I shared a bedroom with Godmother and her mother.

The secretary, Mrs. Turvey, was a pleasant young lady, very helpful to everybody, and who didn't seem to mind being called Topsy. She assigned my godmother, Babushka and me to a room in Block G. This was a good arrangement for me at the time—I still enjoyed freedom of movement with no rigid adult influence or supervision. Like everybody

Relaxing in Baginton Fields at 14 years old, near Coventry. c1946

Coventry Technical College, c1945-46

else, I enjoyed a weekly government cash allowance and my own government-issued rations for clothing and confectionery; the coupons had no cash value. I was also entitled to free schooling and certain school supplies. I didn't have any independent income, of course, so my coupons were mostly no use to me, but others had independent resources, and they coveted coupons from people in my situation. I think my godmother was on salary from her employers in Hong Kong, with headquarters in Belgium. She definitely benefited from having free access to all my unused coupons.

It was also a good arrangement for my godmother. She was not financially responsible for me or obliged to ensure my welfare, and she benefited from the clothing coupons that were of no use to me and any confectionery coupons left over when I ran out of the weekly government pocket money allowance. The manager for meal services was Mrs. Stollock, a Dutch woman who claimed that at one time she was in service to the Dutch Royal Family.

As far as laundry was concerned, I remember only washing my own woollen socks in the sink in the room and drying them on heating pipes from the central heating system going through each room. My godmother must have visited the central laundry facilities, and taken care of my other laundry, since I was living in the same room with her.

The hostel provided temporary accommodation to RAPWIs from all over the Far East: Shanghai, Indonesia, Hong Kong, Singapore, Malaysia, Philippines and places in between, representing many different nations. Many of the people were former Stanley internees. The hostel atmosphere was friendly and optimistic, but not everyone thrived in it.

There was a man from Indonesia in my block who was studying accounting by correspondence from ICS. He rarely ventured out of his room. He seemed frail and unhealthy, and more than once he overmedicated himself to the point of needing urgent medical attention. I realize now that he was probably suffering from post-traumatic stress disorder after his wartime experience in Indonesia.

On the other side of the block where I was living were the Murray-Wilsons, an American family who had been interned by the Japanese in Shanghai: Robert, his wife and their two children. Petal was a toddler still in diapers, and Bobby was about five or six years old. They were a lovely family, and I often played with the children.

I think both my sisters were originally billeted in the same complex as me, but in a different building. I understand that initially Margaret got a job as a stenographer in an office nearby and eventually moved out of the hostel. I know nothing about what Julie did, and I don't think she enrolled in any educational program, so I have no idea whether she graduated formally from school or not.

The social committee at the hostel organized regular dances for young people. I was curious about dancing but didn't have the courage to take part; feeling it was all somehow beyond me, I would make a brief appearance and talk to people I know, then drift away. I often used to see John Barton dancing with his girlfriend Dulcie Lambert, a pretty Dutch girl whose family were RAPWIs (Released Allied Prisoner-of-War Internees) from Indonesia.

One of the first thing I did in Coventry was to register myself in the academic program at Coventry Technical College leading up to London University entrance examinations. By this time, I had missed half a year's schooling at the college through no fault of my own; I just wasn't in Coventry at the time! With having been deprived of three preceding years of formal education as well, I was greatly challenged to keep up with my class. This ultimately meant long hours studying at nights because none of the teachers appeared sympathetic about my backwardness, and quite rightly so. I just had to make it, or I didn't! Not surprisingly, I didn't do well in French, and I surprisingly struggled with English. Nevertheless, I was confident that I would overcome these deficiencies in the final year.

Even though I enjoyed studying school geometry, I well remember when Corrine, a fellow student, and I were detained after classes to memorize Pythagoras theorem flawlessly. Another student in my class was Eddie Eleazer, who, with his family, was also staying at the Baginton Field Hostel. We both enjoyed the English history course and had a lot of fun quizzing each other on historical events and dates; this really sharpened our knowledge at exam time. Eddie and I would meet up again in Hong Kong the following year. His brother Moses was a member of the HKVDC who had been incarcerated in the Sham Shui Po Military Camp in Hong Kong during the Japanese occupation.

Despite challenges, I enjoyed my time in school and have fond memories of riding double-decker buses to and from school with my classmates, sometimes scrambling to finish homework; of visits to the college tuck shop across the street; of purchasing my first book—a

Concise Oxford Dictionary, which I still have in my possession; and of taking evening classes in introductory Russian, at my godmother's suggestion, in deference to my Russian mother's heritage. Because of my lack of interest in the Russian language, I never finished the course.

On the trip home from school, we got off the bus at the Chace Hotel and from there walked about a quarter of a mile back to the hostel, passing by Farmer Mills' horse pasture. Two horses, named Flicksy and Flocksy, would often canter up to greet us at the wooden fence. In season, we fed them apples from one of the farm's Cox's Orange Pippin apple tree.

§

Mr. and Mrs. Murray-Wilson knew that I enjoyed amusing their two children, and the children seemed to enjoy my company. One day, they asked me if I would mind babysitting the children for a whole Saturday while they drive off to Birmingham on a shopping spree. I didn't know anything about babysitting, but without hesitation, I agreed.

So, come the appointed Saturday morning, the two children were dropped off at my place. My godmother and her mother were quite amused and wondered how I would manage. Well, to tell the truth, I wouldn't have managed if my godmother didn't help me out occasionally at key moments, especially with diaper changes for Petal.

At the end of the day, I was glad to see their parents and was able to report that everything went well. They had a discussion with my godmother and seemed pleased. As far as I was concerned, I'd done my good deed, and compensation never entered my mind.

After school next Friday, Mr. Murray-Wilson told me to meet him after Saturday matinee at the Hippodrome Cinema in town centre, very near the site of the cathedral that was demolished by German bombs during the war. Imagine my surprise when he led me into the Halford Bicycle Shop nearby, and turning to me, he said, "Pick whatever bicycle you like."

I was so stunned that I lost my voice. He had to repeat what he just told me. Then he gently helped me choose one of the bikes. The final choice was a Hercules, equipped with dynamo-activated head lamp, a bell, a Sturmey Archer 3 speed gear and a rack with a bag at the rear. Can anybody imagine how I felt to receive such a valuable gift after so many years of next to nothing? I don't even know if I adequately thanked Mr. Murray-Wilson. This bicycle would be my prized possession for all the coming years ahead, up to the time that I set off into the world on my own.

I took great pleasure in riding my bike the three miles to school and back rather than taking the bus. The rear bag was handy for my books, and the stand at the back was useful to give friends short rides now and then. I polished my bike daily, including every spoke of both wheels. I became adept at repairing punctures, and always carried a patch kit for the inner tubes with me. The wrench for the bike had four hexagonal holes of various sizes at each end, suitable for all the nuts on the bike.

I became a de facto member of a small, exclusive group of bicycle owners at the hostel who flaunted our mobility at every opportunity. We rode our bikes from the accommodation blocks to the central building for meals, swapped stories about our cycling prowess and even demonstrated for admiring (or perhaps envious) non-owners.

One day, I was freewheeling along at a good clip when a young boy suddenly sprang out from nowhere and thrust a wooden stick into my front wheel before I could avoid him or even realize what he was doing. The stick caught against the front forks, bringing the wheel to an abrupt halt. I flew over the handlebars and my bike crashed to the ground.

"My beautiful bike!" I cried. Without checking for my own injuries, I hurriedly examined my wounded Hercules, lying lifeless on the ground, thinking surely it would be bashed beyond repair! After calming down, I could see that the frame and fenders were undamaged, but several spokes in the front wheel were badly twisted. That was more than I could repair, so I took the wheel off and had it fixed up in the bicycle shop. After that, my Hercules was no worse off than it was before the incident. I never did see the boy again.

Weekends and holidays were the best days for riding together with friends like John and Leo Barton. We cycled to Leamington Spa on one occasion and to Stratford-upon-Avon on another. Ashby-de-la-Zouch, a pretty market town in Leicestershire, was a pleasure to visit. Those were glorious days.

§

In the spring and summer of 1946, my godmother took me with her on several coach excursions around the countryside. I particularly enjoyed visiting the magnificent Warwick Castle and its grounds. Warwick Castle was preserved because the Earl had sided with

Cromwell and his Round Heads, but neighbouring Kenilworth Castle was razed to the ground. Stratford-upon-Avon was also visited.

We also made a trip to London in the summer, and there revisited some sights we were familiar with: Buckingham Palace and the guards were still impressive; Trafalgar Square with all the pigeons flying about; Madame Tussaud's Museum was fascinating; and a stroll through Hyde Park was interesting, especially the ducks in the ponds. Most importantly, we picked up our new passports. I still have a copy of this passport.

Living for the present only, and with no thought for the future, I was very contented and happy during this phase of my life. The freedom with no responsibilities was an extra bonus. My godmother too was very agreeable towards me, which added to my contentment.

In keeping with my nature at the time, I wanted to share my happiness with my godmother, the person closest to me at the time. I sold a part of my stamp collection to a stamp dealer, and used the proceeds to buy her some chocolates, her favourite treat. She had no trouble accepting my gift, and she even shared some of it with me. I made her happy, and that added to my own happiness.

§

During summer holidays, I got a temporary job at Armstrong Siddley motor works in Coventry. They produced beautiful cars, which I think sold for £1,000 each at the time (equivalent to about £38,000 today). I was assigned to the finishing department, where several of us juvenile workers stitched leather upholstery onto car seat frames. We sometimes lost the curved needles in the process, and if we couldn't retrieve them, we'd just got another one and continued with our work. Test drivers sometimes complained of being 'stuck with a pointer' when they were testing the brakes with force. We boys were being paid for piece-work, and I never complained about the half-crown I received for each seat completed.

§

Margaret eventually married James Jackson, a soldier she met on *Highland Monarch* whose home was in Birmingham. James was a commando, a member of a Special Force fighting behind Japanese lines in Burma, and one of the soldiers that I had much admired. He had told harrowing tales of his experiences in Burma during the war.

I have no idea when Margaret moved out of the hostel, and indeed I only learned about her marriage after it had taken place in the Birmingham suburb of Oldbury on June 29, 1946. I have a group photograph of the wedding party, with appropriate wording addressed to me from Jimmy, but no comment from my sister; the photograph showed Julie in the group. Why wasn't I invited? My relationship with my sisters had always been fragile; clearly, that had not changed. Nor would it.

Julie was in residence at the hostel, but here again there was very little contact between us. She wasn't attending school, so I suspect she must have found a temporary job in the city.

§

England was beginning to feel like home. I loved everything about England: the people, the countryside, the towns and cities, the history and the freedom. One of my favourite subjects in school was English History from 1688 to 1914. I never thought for a moment that soon I would be forced to give up this idyllic life.

Rehabilitation Program Ends

In September 1946, the Ministry of Health announced to all RAPWIs at the Baginton Fields Hostel that their guest status in the UK was coming to an end, and it was time to consider choices for their own future. Those with relatives in the United Kingdom had moved out already, and the remaining hostel residents were offered two options:

1. Return to their place of origin. The Ministry of Health would arrange for transportation to the port or city nearest the desired destination, in two months' time, which would be November.
2. Start a new life on their own in the UK with government assistance.

My heart sank; this indeed was a jolt! My formal education was incomplete—I had just commenced the final year of the London Matriculation program at Coventry Technical College. What now? Of course, there were numerous government assistant programs for a young person in my predicament, only I wasn't aware of them. I was scared and didn't know what to think or do. Godmother didn't offer any helpful suggestions or advice.

§

My godmother decided to return to Hong Kong, where her prewar job and position awaited her. Margaret had been recently married and was now living in Birmingham with her husband. Julie elected to return to Hong Kong, where she was going to fend for herself and look for a job. I was the only wild card.

Margaret, being the oldest surviving family member, had the legal right to take care of me, and benefit from government assistance. That was the most logical course of action that could be expected and could have led to a significant moment in my life!

Margaret and Godmother met with me. "Do you want to remain in England, or do you want to return to Hong Kong?"

I was 14 years old so being on my own was not an option. Regarding Margaret, the following thoughts were crowding my mind: she was newly married—surely a brother in the same house would be most inconvenient? I had no reason to believe in her affection for me, considering so little had been revealed to me from her all my life. As far as my godmother was concerned, I had a real fear of her. These were all real concerns.

Margaret expressed no positive interest in my welfare, nor did she offer any suggestions at this meeting. Faced with no options, I remained mute, uncertain how to respond to the two choices! Finally, Godmother addressed me directly, without looking at Margaret. "Robert, you can come back with me to Hong Kong." Apart from this implied commitment, she didn't make any promises.

With no response or counter offer from Margaret, I accepted Godmother's offer, without knowing what plans were in store for me. I later wondered why Margaret didn't take me aside to talk to me.

Many years later, she remembered this meeting, and asked me why I didn't speak up. My response was that she had made me no offer of any kind; she herself had remained silent. To this day, I am satisfied that I made the right choice, as it was with Godmother's support that I successfully launched myself into a marine engineering career.

Departure for Hong Kong

On November 5, 1946, Olga, Babushka, Julie and I travelled by train from Coventry to Her Majesty's Naval Base at Plymouth, and boarded HMS *Victorious*. She was a Formidable-class aircraft carrier that had been modified to transport troops, supplies and civilian personnel between the UK and places in the Far East and Australia. The complement of the ship had been greatly reduced for peace-time voyages, which freed up regular crew accommodation for civilian passengers.

Shown on the passenger list was a Mrs. O Robinson (my godmother), with two children (unnamed), obviously my sister Julie and me. Upon arrival into Hong Kong, Julie would initially stay with Costia and Lydia Kriloff and then go on her own.

A dozen boys, varying in age from 10 to 18, were billeted in a single mess with bunk beds. I was assigned an upper bunk. Brothers Ian and Murdoch Nicholson were also in upper bunks next to me. Murdoch was a very restless sleeper, and I often had to push him away from encroaching onto my bunk in his sleep.

All passengers received the pamphlet "HMS *Victorious*: A Guide for Civilian Passengers", detailing the facilities available on board the ship and outlining 'dos and don'ts' for everyone's comfort and safety. The ship facilities included laundry, church, hairdressers, cinema, shop and a library "situated in the upper level adjacent to the chapel and the Torpedo Body Room." Leather craft and felt materials were provided to create toys.

It was a thrill to be travelling on a real aircraft carrier with aircraft tethered to the flight deck and below in the hangar deck. The engineers even allowed a visit into the engine room; I understood only a little about the machinery but found it impressive and awesome. The machinery space, including the boiler area, were under constant air pressure. Entering and exiting the pressurized machinery spaces involved passing through a double-doored air-lock chamber. This design feature improved combustion in the boilers. This experience impressed me further to consider seriously a career in the navy, perhaps as an engineer.

The voyage took a month, and the ship called at Gibraltar, Malta, Port Said, Suez, Aden, Colombo, Trincomalee and Singapore, before finally arriving in Hong Kong on December 5th. Each time the aircraft carrier entered a port, its pennants fluttering, the Royal Marine Band would play as it marched up and down the flight deck. Dignitaries would come on board ship and be ceremoniously welcomed. Watching it all was rather intoxicating.

Trincomalee, in the Eastern Province of the island of Ceylon (now Sri Lanka), is where all the Seafires (naval version of the famous Spitfires) were offloaded onto barges, which were then delivered to the RAF base on the island. The off-loading operation was fascinating to watch.

The ship's complement also organized activities to entertain the civilian passengers, including competitive sports (I have long since lost the first-place trophy for my 13-second 100-yard dash). Most nights, movies were shown in the hangar, on a screen mounted onto the deck of the elevator and raised to a comfortable height. A boxing ring was also set up on the elevator deck, and featured bouts between crew members, and anybody else who wanted to demonstrate their prowess at boxing.

I couldn't help but feel a twinge within me as HMS *Victorious* steamed through Lye Mun Pass into Hong Kong's harbour early in the morning of December 5. I wasn't unhappy to be back in familiar surroundings, but the freedom I had enjoyed for the last 14 months was, I knew, at an end. The future was uncertain; all I knew for now was that I would be living with my godmother. For how long, I had no idea, but I was acutely conscious of the dependence in which I would be living under for the foreseeable future.

HMS *Victorious* anchored in mid-harbour, and the Royal Marine Band gave their usual stirring performance, marching up and down the flight deck as the passengers disembarked into launches with their luggage, to Queen's Pier for those going to Hong Kong Island, or to Police Pier for those who were going to Kowloon.

My treasure was delivered to my address a few days after our arrival. I did not realize how concerned I had been about my bicycle until I saw, with great relief, that it had made the voyage safely.

A group photo taken on HMS *Victorious* on its way to Hong Kong. There were civilian passengers and military personnel on board on this voyage. Included in the photo are Claude and Sydney Hollands and me (on the front left). c1946

HMS *Victorious* at anchor in Hong Kong Harbour upon arrival. c1946

CHAPTER 10: SELF-DISCOVERY AND ENDURANCE

Adolescence in Hong Kong

1946 – 1952

Godmother worked in the accounting department of the Crédit Foncier d'Extrême-Orient, a Belgian real estate firm that owned extensive properties in the Far East, including in Hong Kong. Godmother was eligible for subsidized staff accommodation, and we moved into a corner flat on the second floor at 220 Prince Edward Road, Kowloon. This was to be my home for the next six years.

The flat consisted of two bedrooms, one bathroom between the two bedrooms, a large living room with a coal fireplace (never used), a small dining area and a kitchen in the back adjacent to the servants' quarters. The wrap-around verandah overlooked Prince Edward Road and the KCR railway bridge across the side street. The floors were all teak boards except for glazed ceramic tiles on the verandah and in the bathroom. Mounted over the bathtub was a copper-sheathed, gas-fired hot water geyser with brass fixtures.

To accommodate me, the small dining room became my sleeping area; there wasn't much privacy, as it was in the pathway between other rooms in the flat. There was a window which looked down on a courtyard at the back of the flat; it was part of the flat occupied by the property caretaker and his family. The extra-large front hall space was also used as a dining area, which was convenient, as it was close to the kitchen. The two bedrooms both opened out to the verandah.

The caretaker in the flat below us owned a chow dog, which was often tethered to a drain pipe. My godmother encouraged me to take the dog out for walks, and this soon became a normal routine which provided me an escape from the flat. The dog and I would roam around the neighbourhood, through the high-end residential areas along Kadoorie Avenue, St. George's Crescent and Braga Circuit, or along the railway track embankment. The dog and I bonded well; it would caper joyously when I came to take it out, and whine for attention at the sight of me at the window above the courtyard.

§

La Salle College, near our flat, had reopened soon after the end of the war. I enrolled and started classes in January 1947 to complete the schooling I had started in England.

I was delighted to find Eddie Eleazer, a friend from Baginton Fields Hostel in Coventry, in the same classroom. We often worked on homework assignments together and quizzed each other just before tests. As the college was operated by a Catholic institution, I was surprised to see Eddie, whose father was a rabbi. There were practical reasons for the choice: the school had a good academic reputation and was conveniently located close to their home in Kowloon.

Eddie was about six months older than me, which made a significant difference, as far as matriculation exams were concerned. Eddie qualified for his matriculation examinations leading to university entrance in June 1947. To write the matriculation exam, a candidate is required to have attained the age of 16 -- I was short four months.

The principal of the college was Brother Cassian, a French-speaking Belgian. He was quite good to all the students but had a penchant for tweaking the ear of any younger student passing within his reach. Brother Honorious lectured on the Bible (for the Catholic students) and taught geography. Brother Raphael taught English and history and showed us an efficient technique to take down notes during lectures—very useful when attending university. Mr. Xavier, a lay teacher, was a brilliant mathematician who had developed a great technique for solving factorizing problems in algebra. Another lay teacher, a Chinese gentleman whose name I cannot recall, taught us physics and chemistry. He often handled acids carelessly during laboratory classes, and all his clothing seemed to have at least one burn hole somewhere.

§

That summer, I drifted aimlessly, often cycling to King's Park to watch boys playing baseball. I knew many of the players from the school, and when they were short of players, they invited me to participate. I didn't have any equipment of my own, but they waved this off and shared theirs with me. This freedom would soon end, as my godmother found ways to fill my time at home.

Godmother had one servant, Ah Ching. She came from Indo-China (Vietnam) and spoke pidgin French, which was helpful to my godmother, who spoke none of the local Chinese dialects. Ah Ching did most of the cooking and some of the cleaning in the flat. My godmother liked to entertain and was particular about the look of the place before a dinner party or entertaining visitors.

Because Ah Ching was occupied with cooking and other duties, and Godmother claimed to have a heart condition (palpitations, she called it) that prevented her from physical exertion, the cleaning, polishing and maintaining of floors became my responsibility, including other related house-cleaning activities. All tasks involved were performed under Olga's direct supervision.

A major polish of the floor involved clearing all furniture from the area to be waxed, including potted plants and their stands, the aquarium tank, etc. All exposed floorboards had to be swept clean (there were no vacuum cleaners in those days) and then wiped with a damp cloth. This was followed by the application of floor wax and final polishing by a weighted brush. This intensive labour was both tedious and time-consuming. The whole process took several hours, and in the sub-tropical heat and humidity of the spring and summer months, it was also physically exhausting. When the task was completed, I could always look forward to a good rest and uninterrupted reading time; that was my favourite reward!

Other polishing activities included the dinner silverware and the copper geyser shell of the hot water heater in the bathroom. Polishing the copper shell of the hot water geyser in the bathroom was a weekly chore and took an hour with Brasso.

Godmother also had a passion for potted plants. There were plants in the living room, plants in the two bedrooms, plants on the verandah floor and plants on the verandah parapet that stretched two sides of the flat. There were about 20 or more varieties, and they made a nice show in the house. With the threat of a typhoon, all potted plants on the parapet had to be taken down to the floor for fear that some, if not all, would be blown away.

The fish aquarium in the living room also needed my attention; it was about 18 x 24 x 42 inches and was mounted on a stand. Algae control and cleaning was a constant chore, but well worth it to see a nice display of various small tropical fish swimming in the tank.

The chore I never minded, even if it interrupted my reading, was running errands with my bicycle, to pick up an order at Tkachenko's on Hankow Road, or at Cherikoff's on Austin Road. I rarely had to go more than 10 kilometres each way, but it was a pleasure to be out on my bike and out of the house for a few hours. I sometimes took a longer route to delay my return home.

§

My godmother was an avid reader with a passionate interest in historical fiction stories. I read most of the books after she had finished with them, and we quite enjoyed discussing the books with each other. When I had the opportunity, I loved browsing books at the Swindon bookstore on Nathan Road.

Georgette Heyer's historical novels were my favourite at the time, because of my interest in history which she used for her themes. I was also introduced to fascinating new worlds by such authors as Rider Haggard, Robert Louis Stevenson, Rudyard Kipling, Edgar Rice Burroughs, Charles Dickens, H.V. Morton, Winston Churchill, Jules Verne, Daniele Varé, Richmal Crompton, Zane Grey, Somerset Maugham, Agatha Christie, Alexandre Dumas, Rafael Sabatini, Joseph Conrad, Daniel Defoe and many others. Stories of adventure, romance and endurance opened a wonderful world to me and gave me hope that the wide world outside had prospects beyond what I could imagine; all was waiting for me.

§

I liked Uncle Harry. He was as jovial as he had been before the war, despite now using a cane and wearing a prosthetic boot on his right foot because of the injury he had suffered in Stanley Camp. Uncle Harry often came to the flat to visit Godmother, like he used to do before the war, despite the cold shoulder he received from Babushka. I don't think he was totally aware of my lifestyle under Godmother, therefore he had little reason to think I was disadvantaged. Maybe it was because I was so good at concealing my feelings.

Probably at the instigation of Godmother, Harry approached me one day to explain the physical difference between a man and a woman, and what the differences mean. I can remember no details of this conversation and have a feeling that he didn't know how to approach this topic. But what I do remember was hearing him discussing this episode with Godmother afterwards, and how they both laughed at Uncle Harry's awkwardness. So, my education about sex didn't go very far and I remained as naïve as before.

§

I had many occasions to meet Godmother's circle of Russian friends during the early years of my stay with her. I think many had also been friends of my mother and most seemed pleased to see me, but as the conversations were mostly in Russian, my interest was not too great.

However, I did gravitate towards Mrs. Nesteroff whenever she visited. Instinctively, I felt that she was sympathetic towards me just by the way she used to look at me and how she spoke to my godmother about me, even though it was in Russian. I would not be surprised if it was Mrs. Nesteroff who suggested to Godmother that I visit her once a week to learn some Russian language at her home, and that is how it began. By this time, both her sons were at university in Australia.

The lesson sessions started off well, and initially I concentrated on vocabulary. In time, this expanded to mastering simple sentence structures. To promote conversation, Mrs. Nesteroff often encouraged discussing my activities at home. This naturally revealed my current lifestyle. What Mrs. Nesteroff thought about it, I don't know, but she must have known how I felt because she had raised two sons herself. The lessons were eventually terminated suddenly, without any explanation. I certainly missed my visits with Mrs. Nesteroff.

I was always very careful never to share negative feelings I had about living with Godmother; after all, I was dependent on her. However, I had an unusual sympathizer right in Godmother's home—Babushka, her mother. She and my godmother had heated words from time to time, which I felt were about me. It didn't make much difference in my life, but it was nice to know that I had a champion at home. Perhaps the arguments weren't always about me, because Babushka also had an intense dislike towards Uncle Harry and showed this by ignoring Uncle Harry when he came to visit.

§

Godmother was friendly with a French lady who was a tenant in our building; she had a daughter named Monique, who was about my own age. My restricted life allowed little time for socializing, so I hardly knew Monique, but one day, Monique's mother asked my godmother whether I could join them for a day at the beach. I was delighted when Godmother consented to let me go.

On the appointed weekend, we set off in Monique's mother's new Hillman Minx for a beach on Castle Peak Road. When we arrived at our destination, her mother had some difficulty parking the car at the beach parking lot, which was just off the main highway. In attempting to avoid the heavy flow of nearby traffic, she accidentally stepped hard on the accelerator when it was still in reverse and the car landed on the beach upside down with the three of us scrambled up inside. Fortunately, with the assistance of people from the beach, we got out of the car without any injuries. I think Monique and I took the bus back home, and her mother stayed behind to deal with towing her car to a garage.

§

The Tipples were a family that Godmother and Uncle Harry knew from before the war, and they remained good friends over the years. I had been introduced to them at their home near Kai Tak years earlier. Red Roofs was the name of the house. Mr. Tipple had passed away in an accident in Calcutta early in the war. But Dorothy Tipple and her two children had returned from Australia, where they had been evacuated in 1940. Now they were back and living in an apartment on Waterloo Road, not far from where I was living.

I didn't see much of Berry, Lesley's brother, but she and I did see each other from time to time. I remember meeting Lesley to go for bicycle rides in the neighbouring area, along Nathan Road to Prince Edward Road, then at the intersection of Waterloo Road. I always felt welcome when I cycled over to visit them in their comfortable HSBC quarters on Waterloo Road. I also remember their Cocker Spaniel dog, Rogan. Those were fun visits.

Not long after their arrival back to Hong Kong, Berry commenced his marine engineering apprenticeship at Taikoo Dockyard in Causeway Bay, which was quite a commute from his home. It was also at this time that Godmother started giving thought to my own future.

§

I had lost track of Julie after we disembarked from HMS *Victorious* in Hong Kong. I later learned that she lived with Uncle Costia and Aunty Lydia for a short time, and then shared a rental apartment with some of her friends, including a university student she had met in England. She was only 16 when we landed back in Hong

Kong, and she too had nobody to care for her, so had to be resourceful on her own. I believe she completed a secretarial course to enhance her opportunities for employment.

I only remember Julie visiting me once at the Prince Edward Road flat. It wasn't a pleasant visit, because my godmother was unfriendly towards her. I have no idea why she disliked Julie so much, nor did I dare ask. But when the visit came to an end, she told Julie in no uncertain terms never to visit again. The lack of a friendly welcome there is probably one reason I rarely saw Julie while I was living with my godmother.

I have no recollection of seeing Mrs. Kovac in my life from the past, but she definitely knew Mother, and perhaps even Father at one time. This generous lady had gone to the expense of installing a monument at Mother's gravesite in Happy Valley, about 10 years after Mother's burial. Julie joined me at the gravesite, together with Mrs. Kovac, to pay respect to Mother's memory.

gwulo.com

Prince Edward Road looking west from RR Overpass. For six years I would be cycling all along this road part of the time to La Salle College (visible) and HWD. I would also walk to Sunday church service at St. Teresa's Church (visible). c1940s.

The flat at 220 Prince Edward Road where I lived for six years with Godmother Olga. It was decorated for Christmas; note plants and well kept hard wood floors. c1946.

Godmother and Harry, postwar in Kowloon

Portrait of my godmother, Olga Robinson, taken in Hong Kong. c1947

Apprenticeship at Hong Kong & Whampoa Docks (HWD)

There were several considerations that led me to enter an apprenticeship. Firstly, due to my fractured formal education between the ages of 7 and 15, the path towards higher education was not going to be easy, if that was to be a chosen objective. Even if I qualified for higher education, I had no access to financial resources to accomplish this objective.

Also, at 16 years of age, I was aware of my own vulnerability and undefined career prospects but had nobody to express these concerns and seek advice. It was a great relief for me when the following decision was made.

Taking a cue from Dorothy Tipple, who had her son start an apprenticeship program, Godmother's good friend from pre-war days, Alec Black, was instrumental in sponsoring me in the same career path, but at Hong Kong & Whampoa Docks. As the principal surveyor for Bureau Veritas, he was well-known in local engineering circles.

For Godmother, this would mean she would have to support me for at least another four years, at the end of which time she would discharge any assumed obligations for my welfare. Unfortunately, she never discussed any plans she might have been considering for me, and therefore I was totally unaware of her positive contributions towards my future career.

With diligent application and success at future Ministry of Transport (MOT) examinations, this was expected to launch me into a gainful engineering career. I was to commence this apprenticeship program on February 2, 1948.

I had no idea what I was getting into, not having any mechanical background or recognizable inclinations in that direction. However, it seemed to me to be a worthy career path; look at the success of Alec Black! My brief exposure to ships to date were the naval ships that I had visited and travelled on immediately after the war, and at that time, I was impressed.

On the morning of February 2, 1948, now 16 years and three and a half months old, I rode my trusty bicycle to the dockyard in Hung Hom and checked in at the main office. I was instructed how to record the hours I worked and then sent to report to Mr. Conway in the engine shop.

All employees, including apprentices, supplied their own work clothes and work shoes. Safety equipment was hardly thought of back then; no one wore a hard hat, steel-toed boots, gloves, ear or eye protection. But the employer provided tools—Mr. Conroy gave me a requisition chit and sent me to Stores, where I was issued a hammer and a coarse rasp.

When I returned, Mr. Conroy took me to a work bench, handed me a length of round, solid steel rod, one inch in diameter, with the instruction to file a one-inch square at the end of the rod. This was a painful learning assignment because he rejected my every effort. Each rejection had to be cut off with a hacksaw and the process restarted.

At lunch time, I met a few of the other apprentices, and I learned to my delight that John and Leo Barton, who I had met in Baginton Fields in England, were also apprenticing at HWD. John had already been here a year, and Leo was two years ahead and was in the drawing office by then; both could be considered veterans compared to a rookie like me. I quickly reconnected with them both.

After lunch, I went back to the engine shop to satisfy Mr. Conway that I would be able to develop the skill of filing the end of a one-inch rod into the shape of a square. I don't remember if I ever did succeed in achieving this.

John and Leo introduced me to the other apprentices, all good people. The names I remember include Ian McKelvie, Alan Badham, René Ruyters, Andy Ruyters, Bill Fallon, Joey Rodrigues, Bob Houghton, Tony Oliviera, Eddy Oliviera, Len Hutton, Peter Thompson, Juan Conception and Reggie Xavier (ex-HKVDC and ex-POW) and "Guwsup" (Dog Flea). Reggie was the most senior apprentice and was preparing to go to sea very shortly at the end of his apprenticeship.

The China Light and Power Generating Station (CLP), next to the Hongkong & Whampoa Docks, had three apprentices who usually joined us on our lunchbreaks in the shipyard. They and two of our apprentices owned motor bikes and liked nothing more than to talk about them. The rest of us were owners of the lowly pedal bikes.

In the engine shop, machinery was brought in from the ships berthed at the shipyard for repairs or overhaul. The shop was engaged in general fitting work in connection with the repair of all marine and other machinery. I spent the first 13 months of my apprenticeship in this shop and learned a lot about marine machinery.

The outside fitters department was where work was done on the ships themselves, either as they float alongside the wharf or inside the dry dock. This work includes the repair, removal and installation of heavy marine engines, both steam and diesel. By this time, John Barton and I were working together as a team under the supervision of George Bolton.

We became familiar with removal of propellers, propeller shafts, rudders and the repair of hulls followed by painting. Under George Bolton, we learned about steam machinery and boilers, steam turbines and auxiliary steam engines. Under Peter Beertson, we worked on main diesel engines, shipyard mobile cranes and the 100-ton hammer-head crane.

Terence Barton, brother to John and Leo, whose apprenticeship was interrupted by the outbreak of war in December 1941, had completed it after the war. It was here that I met him when his ship was in dock for repairs. Talking to Terence gave me renewed inspiration for a career of marine engineering. I did notice that, as the result of an accident during his work, he had lost the tip of his middle finger on his right hand.

John and I spent a total of 23 months gaining experience on the different types of machinery used by ships. From here, John Barton went to the drawing office to round off his apprenticeship experience, while I continued training elsewhere in the shipyard.

The brass shop was a machine shop operation equipped with numerous lathes, drill presses and boring machines. Here I learned how to machine screw threads of different sizes and shape and do metal work on brass material; all most interesting. In my three months in this shop, I also fabricated two brass, three-piece candle-holders. I presented them to my godmother as a gift.

In my nine months in the drawing and design office, working under Chief Draftsman Mr. Davidson and George Wallet, I learned basic naval architecture nomenclature, how to read machinery and layout drawings, draw piping layouts, prepare equipment design take-offs for drafting and draw machinery components. Experience in the drawing office was the last stage of apprenticeship.

§

It was during my apprenticeship on August 11, 1949, that I was lucky to witness the famous HMS *Amethyst* proudly steam into Hong

Kong's harbour with all flags flying! Her hull was badly rusted, and the ship exhibited recent battle damage.

What made the frigate HMS *Amethyst* famous? She had just escaped a blockade by the communists on the Yangtse Kiang River, where she had been held captive since April 20, 1949, a period of nearly four months. After making a successful dash down river and being met by the destroyer HMS *Concord* at Woosung, she steamed into Hong Kong's harbour on August 11, 1949. Almost the entire dockyard personnel, and indeed the rest of Hong Kong, watched her steam past with pennants flying. All the tugboats and ships in harbour gave her a riotous welcome, hooting their whistles. I was thrilled to witness this historic occasion!

§

One afternoon, George Bolton asked John Barton and me if we would like to go out on the tugboat *Pauline*, which was leaving to aid a merchant ship in distress outside Lye Mun Pass. Naturally, we both jumped at the chance.

The *Pauline* arrived at the distressed freighter to discover that assistance wasn't needed but stood by for several hours just in case. Pauline was wallowing in a swell close to the freighter, and that rolling made me feel seasick. When *Pauline* finally arrived back in the dockyard well after quitting time, around 6:30 p.m. and still daylight; I was decidedly seasick!

I cycled home, feeling quite wretched, only to find myself locked out and Godmother and Babushka, both at home, unresponsive to my ringing the doorbell or knocking on the door. Perhaps my godmother was angry that I had not let her know I would be late to come home? She could not have guessed that I could have been out on a ship in the harbour, with no means of letting her know. Her denying me the chance of explaining my lateness was, I felt, very unkind.

Feeling rather sorry for myself, I went around to the caretaker's courtyard and climbed up the drainpipe to get into my bedroom through the open window. There was no response from within the house, so I crawled straight into bed, still queasy from the afternoon experience and slept until it was time to leave for work the next morning. No one mentioned my lateness the next day, or asked how I had got in; I thought it wisest to say nothing.

§

While I apprenticed at the shipyard, I also enrolled in a three-year course in Mechanical Engineering (evening department) at Hong Kong Technical College, leading to a Certificate in Engineering. The classes were held in Causeway Bay, so a bus and ferry ride across the harbour had to be taken. Course subjects I completed included: Applied Heat, Electrotechnology, Applied Mechanics, and Technical Drawing. I received the certificate at the end of the third session, signed by the principal of the school and the director of education.

Len Hutton, John Barton, and me at HWD on board the *Pedro Nunes*, a Portuguese warship. c1950

Group photo of apprentices in front of HWD Head Offices. Back row first two unknown, Bob Tatz, Allan Badham, Tony Oliviera, Leo Barton, Peter Thompson, Ian McKelvie. Front row: Bobby Houghton, McGinnis, Joey Rodrigues, Gow Sup, and unknown. Photo probably by John Barton. c1948.

HKAAF Volunteer Experience

It was during the last year of my apprenticeship at HWD that I joined the newly formed Hong Kong Auxiliary Air Force (HKAAF) as a volunteer. I enlisted as an Aircraftsman 2nd Class on June 13, 1951 as an engine mechanic (air-cooled). I felt proud to be in uniform, a boyhood dream, and to be able to serve my community and my country. Currently, the civil war in China was very intense, with the communists intent on defeating Chiang Kai Shek, the president of the Republic of China.

In the early days of operation, HKAAF was short of pilots, as well as technical service personnel, also known as engine characters (EC). Because of this shortage, these two groups were spared basic training activities during camp and flying maneuvers from Kai Tak HQ. For the rest of us, basic training at Kai Tak headquarters included plenty of formation marching, which we called 'square-bashing', to get us parade-ready—the military love a parade! There was a lot of competition between the three volunteer services to put on a good show at public parades: air, naval and ground.

We also trained in gunnery and care of weapons, mainly with the Lee Enfield rifle and the Bren Gun. At the Annual Bisley Inter-Service Shooting Competition at the Bowen Road shooting range, I scored well and came out as a First-Class Rifleman.

All training activities were led by RAF ground personnel assigned to the HKAAF. Our senior NCO was Harry Chapple, and under him were three sergeants: O'Connor was our electrical expert, Long our instrument expert and Parker our mechanical expert. They were all great chaps and fun to work with. Parker whistled the tune from the "Mockingbird Trill" every morning as we went about our work; I often hum this tune in the shower to this day.

I had limited contact with the pilots. My role was to assist them into and out of the seat of their aircraft, help them with their safety seat harness, and marshalling to help pilots park their aircraft on the dispersal apron. Other than Scotty Munro, who took me up for a joy ride in a Harvard, and one other who took me up in an Auster, I had little interaction with the pilots.

I don't recall the name of the Auster's pilot, but I remember the aerobatics he put the plane—and his passenger (me)—through, twisting and banking over the junks in Kowloon Bay. The aircraft's two doors had been removed, and the view through the openings to the 'unstable' horizon was more than my stomach could endure. I had to clean up the exterior on my side of the aircraft when we were back on solid ground.

Following is a partial list of civilian volunteers with whom I had frequent contact, either at HQ or in Annual Camp. The Tipples were friends, and those with HWD were workplace colleagues: Blumenthal (EC), G. Wallett (EC) (HWD), I. McKelvie (P) (HWD), B. Tipple (EC) (F), L. Tipple (WA), J. Rodrigues (HWD), R. Ruyters (EC) (HWD), G. Randall (P), L. Mose (P), Munro (P), Marsh (P), A. Badham (EC) (HWD), Peter Thompson (EC) (HWD). [EC – engine character; WA – women's auxiliary; P – pilot; HWD – Hong Kong & Whampoa Docks].

In October 1951, I attended Sek Kong Annual Training Camp. This is a time when all sections of the Auxiliary Air Force could gather to work as a cohesive unit. The duration was subject to availability of volunteers to get time away from their place of employment. In this case, I believe, it was about 14 days.

One weekend I was on sentry duty guarding aircraft. which were securely tied down in the dispersal area. It was a clear night with the moon shining brightly. Across the tarmac, I could hear the RAF personnel in their camp enjoying a great party, and no doubt some personnel from our camp were also there, especially the ladies. Sometime after midnight, I was watching two personnel staggering across the tarmac toward me, standing in the shadows underneath one of the Auster aircraft wings. These two chaps had great difficulty walking straight, no doubt due to the influence of excessive alcohol. The only reason they were still on their feet was that they were supporting each other.

I was a little alarmed, as I wasn't sure how to deal with these two staggering effigies; my Lee Enfield rifle was just for show, as I was without ammunition -- not that I would have thought of using it anyway. When close enough, I gave the order to halt, in as firm a voice as I could muster; this was promptly ignored. I tried once more with equal success, when with great relief, I saw a two-man RAF Military Police patrol with a dog approaching us; they resolved the situation by firmly escorting the two men back to their quarters.

I witnessed an interesting incident one morning. I had just finished refuelling one of the Harvards and was busy shooting some photographs with my camera, when the alarm went out that one of our Auster airplanes with a trainee pilot found himself executing 'subsequent actions' (the standard procedure when loss of power in flight occurs) instead of 'simulating a practice engine failure'. Sure enough, standing on the wing of the Harvard, I could see one of our Auster aircraft coming to an abrupt stop in the middle of a rice paddy field, ending up with its tail in the air. Fortunately, there were no injuries to the instructor or trainee pilot, but the Auster aircraft was a write-off. I think the engine was salvaged for spare parts.

The normal procedure for simulating a practice engine failure when flying is to only shut off fuel at the throttle. In this incident, without the instructor's knowledge, the cadet pilot had forgotten that he had also shut off the main fuel valve located alongside his seat. When attempts were made to re-start the engine from the throttle, nothing happened. As the aircraft was too low to have time for recovery, the simulated exercise turned into a real emergency.

On another occasion, an Auster had to come in on a 'subsequent action'. When the aircraft engine was turned off after parking at the dispersal area, fuel immediately started to gush out from one of the two fuel pumps; the sediment bowl from one of the fuel pumps had detached, and of course, the engine was starved of fuel.

My dependency on my godmother for shelter and sustenance was rapidly coming to an end, which would be effective on March 27, 1952. For the entire four years that I was on payroll, I handed the entire unopened envelope to my godmother. After two pays, Godmother would leave me $10. This gesture of mine was to contribute to my upkeep and acknowledge any generosity extended to me. Many a time, at Godmother's request, I took Babushka to foreign movies at the Mong Kok Cinema, which was close by. I usually paid for her ticket and mine with my $10. These movies were in Russian, which was entirely useless to me.

My effective date of resignation from the Hong Kong Auxiliary Air Force is recorded as April 8, 1952, having attained the rank of Leading Aircraftsman (LAC) in the trade of Engine Mechanic. In fact, my activity in the service ceased as soon as I joined the merchant navy.

Gordon Randall getting ready for training flight in Harvard. c1951.

Bob perched inside Harvard in summer camp. c1951

Bob and Auster Aircraft at Summer Camp. c1951

Bob Tatz at HQ on Joey Rodrigues' motor bike. c1951

Bob, LAC (Leading Aircraftsman). c1951

CHAPTER 11: PROMISING FUTURE

Jardines First Contract

1952 – 1955

Without any idea of how to approach a prospective employer, I presented myself without an appointment at the offices of Indo-China Steam Navigation Company, located in Alexander House in Hong Kong, and asked to speak to Mr. John Green, the Superintendent Engineer.

Whether this was considered unusual I did not know, but Mr. Green agreed to see me. I informed him that I was seeking employment as a marine engineer. He looked dubiously at me; with my short stature, slight build and beardless face that needed shaving only a few times a month, I looked more like a 17-year-old than my actual 20. I also informed him that I had completed a four-year apprenticeship program at HWD. In that case, he suggested, I must apply to the Government Marine Department to write Part A of the MOT Second Engineer's Examination. I was welcome to report back to him when I had passed the examination.

I obtained a certificate of service from HWD on February 9, 1952, following which I applied to write the examination. By the end of the month, I had passed the examination and promptly presented myself at Mr. Green's office again. He was surprised to see me so soon. I handed him the certification confirming that I had successfully passed Part A of the Second-Class MOT examination. Surprised and amused, he congratulated me, and told me to stand by for further instructions from his office.

To this day, I cannot recall what my starting salary with Jardine Matheson was going to be. All that mattered to me was that, because I was signed on as an expatriate, I would be entitled to six months paid home leave in the UK after three and a half years of service. I instructed the company to deposit my monthly salary directly to a bank account in the UK, which I would open specifically for this purpose. The monthly high cost of living allowance (HCL) was to be paid directly to my bank account in Hong Kong.

As I left this lucky interview with Mr. Green, I could feel one chapter of my life closing and another one opening.

Jardines – The Company

Jardine, Matheson & Co. Ltd (Jardines), known in Chinese as EWO (怡 和 洋 行), meaning 'happy harmony', was founded by two Scotsmen, Dr. William Jardine and James Matheson, in 1832. They were both former employees of the East India Company, whose trading monopoly in China was coming to an end.

When the East India Company monopoly ended in 1833, merchants and ship owners in England began sending cargoes to the Agency Houses in Canton — especially to Jardine Matheson. The firm had connections in India and throughout the East, a large cash flow derived from their trade in opium and comparatively modest capital commitments. They provided and controlled many banking facilities in Canton.

With increasing volumes of goods to be moved, transportation became an important element in the business model, and Jardine, Matheson & Co. Ltd. entered the shipping business and met with success.

The Indo-China Steam Navigation Co. Ltd. (ICSN) was founded in London in 1881 by Jardines to run steamers on the coast and rivers of China to the Straits Settlements and Calcutta. The operational head office was quartered at Jardines' headquarters in Hong Kong. This initial arrangement fixed Jardine, Matheson & Co. as perpetual general managers of the shipping company and its operation.

Tidying Up Affairs with Godmother

One of the first things I did after returning to Godmother's flat with the good news was to order my new regulation merchant navy uniform, with Uncle Harry's help, from Wo Fat Tailors, who also supplied the cap. The custom-tailored uniform was made from black doeskin material and sported a single gold braid with a purple backing on the sleeves. The wearing of the merchant navy uniform made me feel that I belonged to a worthy community of professional marine engineers and gentlemen officers.

Up to now, I had surrendered over four years of apprenticeship wages to my godmother. Nevertheless, now that I would be fully employed and would soon no longer be living with her, I decided to arrange a monthly bank remittance of HK$200 (£40) to her bank account, as a gesture of gratitude for six years' accommodation and sustenance.

I had very little in the way of clothing or possessions to pack in my suitcase. The most important items were my engineering textbooks. Uncle Harry presented me with a silver-trimmed set of two hairbrushes and a comb, as well as a Rolls razor kit as a going-away gift.

Before I left Godmother's home to join my ship, I went downstairs to the caretaker's quarters. The chow dog went crazy to see me, but I was there to see the caretaker's teenage son, to pass along my most treasured possession, beloved friend and companion for the last seven years, my wonderful Hercules bicycle:

"Here, Steven, is my bike. It is now yours. Take good care of it and enjoy".

He was stupefied and tongue-tied! With tears in his eyes, all he could do was to give me a hug, after which I abruptly left. It brought back memories about my own feelings when I first received this bike as a gift many years ago.

New Adventures Full Speed Ahead

Over the next four years, I would serve on the *Wing Sang, Chun Sang, Eastern Queen, Hew Sang,* and *Wo Sang.* My home would now be my cabin on board my ship, and my mailing address was always c/o Jardines' HQ in Hong Kong.

I had no idea what to expect from my new life on Jardine ships, and travelling to parts of the world with exotic names, and different cultures. My tendency was to wait and see, go with the flow, and be receptive to new experiences. I wasn't going to be disappointed, and the best part of it was that all the officers I sailed with were super guys and they taught me a lot about life and it's pleasures. Furthermore, I was awed with my accommodation, and all the comforts that came with it. I was the new kid on the block!

SS *Wing Sang* 永 生 *(Eternal Growing)*

On Thursday, March 27, 1952, at 20 years old, I joined my first ship, the *Wing Sang*, as fifth engineer, assistant to the second engineer. The *Wing Sang* was one of the last five China Coast cargo/passenger ships of ICSN and was on a liner schedule to Taiwan.

Carrying all my worldly belongings—with room to spare—in one suitcase, I made my way down to the Praya (the promenade by the waterfront) near the Central Market, where the *Wing Sang* was moored.

The area and all the activity were like a page out of a *Terry and the Pirates* comic book (by cartoonist Milton Carniff). With the ship getting ready to leave, the place teemed with people moving on, off and around the ship. Last-minute cargo was loaded, men carrying loads in baskets on shoulder poles scurried and wove among the crowd, crew members brought last-minute supplies on board, passengers visited with friends and family, the quartermaster screened the people trooping across the gangway.

I made my way up the gangway and looked for Paddy Richardson, the chief engineer. After a few pleasantries, he turned me over to Ron Hunt, the second engineer I would be assisting on my first watch duties at sea.

Ron showed me to my cabin and left me there to settle in. Looking around me, I saw what looked like a comfortable bed with two drawers underneath, a writing desk with bookshelves above it and a chair, a cushioned settee with storage under the seat and a coffee table, a full height clothes closet, a wash basin with a mirror above it and running cold and hot water and a porthole to peer out of. Maybe it wasn't much larger than the space I had before, but one thing was for certain—it packed a lot more than what I was used to, and above all, I enjoyed total privacy. This was a luxury I had never imagined, let alone experienced. With great satisfaction, I stowed my few personal effects and went out on deck to observe activities as the ship was being made ready for departure.

Lying in my comfortable bunk the first night out, still savouring the excitement of the day's activities, I wondered whether this was all a dream. But the room and its fixtures were real enough, and I felt eminently satisfied with the world. The ship ploughed through the sea toward Kaohsiung in Taiwan, and the muffled throb of the steam propulsion system and the gentle rocking of the ship lulled me to sleep.

An officer's life on board a Jardine ship was more comfortable than anything I had ever experienced in my life. In addition to the heavenly private accommodations—a sharp contrast to the abjectness of my last 14 years—even junior officers were waited on hand and foot.

There were a few ships in those days where the mere pressing of the service bell in the cabin by the desk would summon a white-coated Chinese steward to your cabin, asking what he could do for you. Cabin service was available until 10 in the evening and started when the ship was in port with a cup of early-morning tea delivered to your bedside. The chief steward, who was also Chinese, oversaw a team of white-robed stewards, called 'boys', who provided all hotel services for the officers and all first-class passengers. The boys cleaned cabins, made up the beds, laundered clothes and served drinks in officers' cabins as well as in public areas. We all lived like gentlemen and the cost was only HK$30 per month.

Paddy Richardson, the chief engineer, was from Belfast. We talked a little about Ireland and his work at Harland & Wolff Shipyards. He was married and had a daughter, Colleen. I would serve under Paddy again later in my career.

The chief engineer does not stand a watch but is available to support his engineering staff on board the ship. In port, there was always one engineer on duty around the clock to respond to service calls or emergencies. At sea, watches in the engine room consisted of four hours on duty followed by eight hours off, around the clock. In port, we worked regular eight-hour shifts, and routine light maintenance work was done from 07:00 to 16:00 hours, with breaks for breakfast and lunch. Only emergency work was done on Saturdays, Sundays and public holidays observed by head office. The second engineer stood watches from 04:00 to 08:00 and from 16:00 to 20:00. These were the hours assigned to me.

All supernumeraries were assigned to assist a watch-keeping engineer. I was assigned to Ron Hunt, the second engineer. The second engineer scheduled all maintenance work on board the ship, directed the engineers and support personnel, whether at sea or in port, and also maintained the ship's supply of consumables for our work. The second engineer was often referred to as a 'tin god' because of his responsibilities and duties. He only reported to the chief engineer.

Ron Hunt, who was the second engineer in charge, was a good engineer, and patiently introduced me to the duties of a watch-keeping engineer, starting with partaking of tea and sandwiches delivered to us at the beginning of our watch. Then we examined the previous engineer's logbook entries. Together, we then patrolled all areas of the engine room, noting condition of all operating equipment. In the boiler room stokehold, I was introduced to the fireman. Then there was the routine of recording pressures and temperatures in the logbook on a regular basis. It wasn't an arduous watch, and before long, it was time to hand over duties to the relieving engineer coming on watch.

It was on the return trip to Hong Kong when, around midnight, the ship ran into some heavy seas, which turned out to be rather unsettling. By the start of my watch at 04:00 hours, *Wing Sang* was bucking a little. I knew something wasn't right with me when I felt no interest in the tea and sandwiches that were served at 04:00. My condition worsened by the hour, exacerbated by the heat in the engine room and the smell of steam and cylinder oil. Ron was understanding and asked little of me on his watch so I could sit down and nurse my misery alone. I was hugely thankful when we finally arrived into port in Hong Kong and the bridge telegraphed 'finished with engines'!

It was a relief when the ship had stopped rolling, and my stomach began to settle. I went up to the exit door at the top of the engine room, and there was Paddy Richardson, grinning at my obvious discomfort, and mischievously extolling the benefits of a big breakfast of fried eggs with greasy bacon, the mere thought of which turned me off any food. Thankfully, it wouldn't be long until I would acquire my sea legs and be able to weather future stormy seas with comparative ease.

It was an unwritten rule in the fleet that serving bachelor officers would relieve married officers of their duties whenever possible when a ship was in their home port. Most officers located their families in Hong Kong because the company's head offices were there, and all ships invariably stopped there for cargo or repairs in the local shipyards. Another favourite home port was Sydney, Australia, but Hong Kong was usually more desirable because the company provided subsidized housing to all married officers and a monthly cost of living allowance (HCL) in addition to their salaries. I had no official residence in Hong Kong or anywhere else; my home was my cabin on board my ship, and my mailing address was always c/o Jardines' HQ in Hong Kong.

Safely back in port in Hong Kong and recovered from my seasickness, I was content to stay on board the ship and act as the duty engineer, either alone or with my fellow bachelor officers, until the ship was to sail again. However, after one trip, on April 5, 1952, I was signed off the *Wing Sang* and was not to sail with the ship again. I was placed on reserve and stayed at the European YMCA in Tsim Sha Tsui, waiting for my next assignment.

For the short time that I was serving on the Wing Sang, the following are the only details I can remember: Year Built: 1938; Gross Tonnage: 3,560; estimated 3,000hp (316nhp); reciprocating steam engine with a turbo-compressor; two oil-fired Scotch marine boilers. Cargo and passenger service. There were cabins for first-class and second-class passengers, and space for 200 to 300 'deck' passengers in the 'tween decks.

SS Chun Sang 祥 生 (Spirit Growing): April 1952 – October 1953

SS *Chun Sang* (spirit growing). As a junior engineer officer, this was one of the earliest ships that I sailed on, and for the longest period. When I signed off the ship, I had completed my full sea-time to write the final Second Class MOT (Ministry of Transport) examination. I was to experience a camaraderie I had never imagined could be so rich and full-filling, and which helped to develop my character and values which has stood me in good stead for the rest of my life.

On April 10, 1952, I was assigned to the *Chun Sang* as the fourth engineer. It was the smallest cargo ship owned by Jardines, operating on a trade route between Calcutta, Straits Settlements, Hong Kong and Japan. It is inevitable that my story about my experiences on this ship will be a lengthy one, because, firstly, it was the most significant start of my next journey in life, and secondly, I served on this ship the longest.

I was eager to learn all about my ship. At sea, my watch-keeping was from 08:00 to 12:00, and from 20:00 to midnight. The *Chun Sang* was designed and built in 1946 in Sweden for Norwegian owners trading in the Baltic Sea and was rated at 2,808 tons. It was driven by a 1,500hp quadruple expansion reciprocating steam engine, with steam supplied by two firetube boilers.

My service on the *Chun Sang* was a very important stage in my development. Not only did I have to learn about operating marine machinery, but also maintenance of marine machinery on board a ship. Because of the lack of design drawings on the ship, I often had to develop drawings of various shipboard piping systems, which of course added to my experience.

My apprenticeship in HWD gave me an appreciation of the technical aspects of marine machinery, but my fledgling maturity was still to be developed. I would have to learn from the following axiom: "You go to school to learn your lesson, and afterwards you are tested; in life, you are tested first, and then you learn your lesson." My fractured formal education would at times create challenges for me.

To me, the most exciting part was operating the main engines at manoeuvring stations, i.e. when the ship is arriving or leaving port. For this, there were usually two engineers, one operating and one assisting.

Having qualified to Part A of the Second-Class MOT examinations at the time I joined Jardines, I wasted no time in beginning to prepare for the final Part B examinations, which would be 18 months from now. I had the momentum, the ambition and the perfect environment to accomplish this goal.

I was to be the longest continuous serving officer on this ship. In 10 months, I was promoted to third engineer and served in that capacity for another eight months, for a total of eighteen months aboard. During this period, there were four changes in chief engineers, two Masters (Captains) and 18 other officers — the richest population of mentors that I could ever ask for!

I made about six voyages on this trade route between Calcutta, Straits Settlements, Hong Kong and Japan on the *Chun Sang*, and visited many ports multiple times. A round-trip voyage from Hong Kong was about three months, depending on the ports of call.

Because the ship had been designed to keep warm air in and cold air out, the ship became a floating sauna for much of the time. We perspired constantly, even when sitting or lying still, and the nearer to the engine room, the worse the heat. The cooler temperatures around Japan in the winter months were always a welcome change.

On hot days, we would rig up a canvas windsock over the engine room skylight in the hope of directing any stray wind into the engine room space. But the ship often moved too slowly to generate any significant breeze, and on days with a tailwind, the windsock made no difference at all. Cabin portholes had attachable metal scoops that jutted out to catch any breeze; most of the time they too were also as ineffective as the windsock for the same reasons. To add to the misery, several of the engineers' cabins were situated atop the boiler room, which meant that the steel floor in their cabins often measured 30 degrees Celsius or more, and electric fans only circulated the warm air. Consumption of salt tablets was quite high among the crew, but I could never swallow a salt tablet without bringing it up, even when I took the tablet with food. The only way I could consume salt was by adding it directly into my food.

Being confined in fairly close quarters for months at a time, it was only natural for officers on the ship to exchange stories about their family upbringing and home lives. Due to my own lack of knowledge at the time, I found it difficult to describe my own origins, other than claiming British citizenship. The only documents in my possession were my British birth certificate, and a British passport issued from London in 1946. Stories about my stormy youth would seem incredible, so I shared none of them. One thing I made clear was that I would be called Bob. Due to my lack of maturity, I listened more than I spoke and learned from the wisdom of mature peers—and folly of the less mature. I observed and then adopted the norms of social interaction, cautious and keen not to be misjudged.

Away from my home port, I seldom ventured ashore on my own; I was always in the company of my shipboard family and assured of returning to the ship for the night, even if some of the veteran seadogs decided to stretch their explorations into the wee hours. I asked no questions and had no idea what they were up to; that part of my education was to come much later.

Notable Memories While Serving on *Chun Sang*

We were at harbour in Hong Kong on my twenty-first birthday; October 11, 1952. I paid a visit to Godmother and Uncle Harry at their home on Prince Edward Road a few days before. They remembered my birthday and very generously presented me with a solid gold Rolex Oyster Perpetual watch. In addition, Godmother also presented me with a suitably-engraved pewter tankard that soon became an essential tool in furthering my education in the very near future. I was very touched by their kindness.

I also received a symbolic chrome-plated key, about eight inches in length, which consisted of a corkscrew, a bottle opener and ice cube hammer, from my sister and brother-in-law. And finally, my shipboard family presented me with a stainless-steel Rolex Tudor watch. I was totally overwhelmed by all this attention and generosity!

All the ship's officers and crew were on a sea-duty schedule because of a typhoon alert. Steam in the boilers was up to pressure and the engines were made ready to manoeuvre if required. My birthday was all the excuse my shipmates needed to throw a rip-roaring party in my cabin in my honour—typical of life on Jardine ships. I was on standby duty from 20:00 hours to midnight.

While everybody else was imbibing alcoholic beverages and having a good time, I was using my new pewter tankard for drinking Coca-Cola. However, during the times that I was briefly away attending in the engine room, my 'worthy friends' had been spiking my Coca-Cola.

At first, there was a spring in my step, contributed to by the camaraderie in my cabin; unknowingly, it was also being fueled by alcohol. As the night wore on, I couldn't understand why I was finding it increasingly difficult to climb up the engine room stairs with the same energy to rejoin the party in my cabin, which was getting more and more boisterous.

When my duties ended at midnight, I was able to fully relax with the company in my cabin and enjoy the celebrations to the fullest. However, it wasn't long afterwards that I succumbed to my condition, at which time my fellow officers tucked me into my bunk and continued the celebrations without my active participation. I woke up the next morning with a whopper of a hangover and a firm resolution never to touch alcohol again in my life. Bill Bennett (the chief engineer) urged me to clamber on top of one of the steam boilers and let the heat 'sweat away' the wretched hangover.

§

Off-duty at sea is usually a quiet time for all the officers on the ship, as they settle into their routine of work, sleep and socializing. I developed a taste for, and skill at, deck tennis, which we played in teams or one-on-one, on top of cargo hatch #2. We only played when the ship was at sea, and at times it would be challenging when seas were rough; throwing against a rising deck and catching a spinning deck quoit in the air required extreme agility and coordination. 'Batchy' West, a great gentleman and mentor to me as well as being the oldest officer on the ship, displayed amazing agility at deck tennis. Alan Noon and Bill Bennett were also worthy challengers playing against me and 'Batchy' West from time to time.

To further my social skills and education, it was here that I was taught the intricacies of certain popular games: Liar's Dice was a favourite, usually played before lunchtime. This required a set of five poker dice, a box to roll the dice from, a bunch of players and jugs of draught. Each player takes a turn at rolling the five dice and challenges the others on the result of the roll. This is where lying comes into the picture, and losers pay for a round of drinks. For the longest time, I was only an observer at these sessions.

What I enjoyed and caught on quickly to was cribbage with a peg board. Two or three players can play this game. Cribbage is really a card game but uses a board to keep score. The object of the game is to be first to score 121 points.

I was very familiar with chess because I used to play this game when I was a teenager in the Refugee Centre in Hong Kong during the war. It's for two players only, and it can get quite intense at times. It certainly helps to while away the time. My partners in chess were usually Bill Bennett and 'Batchy' West. Kibitzing at most of these games was also lots of fun, and some of the wise-cracks were hilarious.

§

The ship was at anchor in Singapore one sweltering afternoon, and Allan Noon, the second engineer; Bill Bennett, the chief engineer; John Stormont, the second mate; 'Batchy' West, the radio officer; Jack Perrin, the first mate; and the third mate Jefferies were all enjoying a leisurely cold beer in the ship's saloon. Then, for some unknown reason,

At dockside in Calcutta preparing to load the vintage taxicab for a destination in Singapore. The two intrepid owners from London are seen here helping with the loading. c1953.

Allan Noon (l) and I wearing our winter rig. He was the Second Engineer, and I was the Fourth Engineer. c1952

Allan Noon & Bill Bennet Deck tennis anyone? Bill Bennett and Allan Noon having a go at sea. Note the match is on covered cargo holds, and netting is fastened to derricks c1952

In Bassein. Two of the guests with three of the officers: Jeffreys, Me as Third Engineer, and the Radio Officer. The uniform we were wearing was known as the "Red Sea Rig". c1953

Shipboard party in Bassein (now known as Pathein), Burma (now known as Myanmar, hosted by the ship's Captain. Ship's officers are tagged, the rest are town dignitaries and their family members, and friends. Ship's Officers: 1. Tony Lapsley, Chief Engineer; 2. Radio Officer; 3. Wong, Fourth Engineer; 4. Wally McNamarra, First Officer; 5. John Stormont, Second Officer; 6. Me, Third Engineer; 7. Captain Lori Cox, Master; 8. Jeffreys, Third Officer. Photo taken by Reggie Xavier, Second Engineer.

Testing a lifeboat and its engine. The four officers are: Bill Bennett, chief engineer; Reggie Xavier, second engineer, Wong, the fourth engineer; and the third officer. c1952

At Sea near Japan. Ship's officers are: "Wally" McNamarra (First Officer), John Stormont (Second Officer), and Allan Noon, Second Engineer. c.1952

Jefferies picked up a jug of ice water and abruptly emptied the jug over my head. Shocked into immobility and gasping from the cold, I could only watch as all hell broke loose around me. 'Batchy' West roared in outrage, chased Jefferies and splattered ketchup all over Jefferies' white summer uniform.

The senior officers quickly put a stop to the ruckus, and fortunately, Jefferies took it well and looked on the incident as a lark, but 'Batchy' West remained incensed at this treatment of me for a long time. No harm was done, and I came out of it feeling none the worse. Later, I asked Jeffries, "Why did you do that to me?" His answer was, "Just for the hell of it, Bob, and I am glad you didn't take umbrage, but 'Batchy' West was sure feisty!"

§

Singapore was always a delight: spotlessly clean, with shops on Orchard Street and Robinson Road, curbside street stalls serving satay, the eye-opening world of transvestites along Boogie Street and, of course, the gloriously colonial Raffles Hotel; favourite waterhole of well-known author and playwright, Somerset Maugham.

We had a shipboard party here, with four or five residents from a ladies' hostel, where one of the officers had a contact. Rae Shawcross and one other lady were journalists, Iris Jantzen was an airline hostess with Malaysian Airlines and another whose name I have forgotten was an office manager.

We were all dressed up smartly in our white summer uniforms and the ladies wore evening finery. The deck was hung with colourful buntings, music played in the background and a delicious buffet was laid out on the boat deck under a clear night sky. The ship's stewards lubricated the socializing with a steady supply of cocktails, visitors and locals got to know each other and future outings in Singapore were discussed or firmly planned. At the end of the delightful evening, several officers escorted the ladies back to their hostel in taxis.

On my next visit to Singapore, I received a message from Iris Janzen, suggesting we get together for an outing. Iris was between flights and the shipping news in the local paper had shown her my ship was in port. We met for lunch and then a shopping spree; it was a pleasant afternoon. Iris was from Ipoh, near Kuala Lumpur, the capital of Malaysia.

Hearing about my date with Iris, Captain Cox cautioned me not to consider any significant commitments. I wasn't sure what he meant, and I didn't think to ask, but my career at sea and Iris' with the airline made our connection fragile at best. I saw her only two other times outside Singapore: once in Georgetown on Penang Island and once in Sandakan in British North Borneo. She eventually married an Englishman who managed a rubber plantation somewhere in Malaysia.

On another visit to Singapore, I met up with Rae Shawcross for lunch and a walk through the Singapore Botanical Gardens. We talked about her hometown in England, her journalistic assignments and her ambitions, and we enjoyed a pleasant afternoon. But the life of a mariner being what it is, I never saw her again.

§

Rangoon, now known as Yangon ('end of strife'), the capital of Burma, was an important port for the economy of Burma. It is situated on the Rangoon River, which is navigable by ocean-going ships. Jardine ships made frequent calls here, and our main cargo was rice for Hong Kong and Japan.

The notable tourist attraction was the Shwedagon Pagoda, built over 2,500 years ago. There is so much gold on the Shwedagon Pagoda that it has been compared to Fort Knox. The structure is entirely gold-plated and studded with diamonds and rubies. The Main Stupa is plated with the gold of 22,000 gold bars, and there is half a ton of gold at the Umbrella. The Shwedagon Pagoda in full view was very impressive.

Allan Noon and I often walked to The Mission to Seamen, a popular drop-in centre for foreign seamen, on our way to the post office or the swimming club. I was appalled by the common sight of what appeared to me to be tuberculosis cases, until I learned that it was chewing of the betel quid (from the nut), that stained the teeth and cavity of the mouth of the chewers a deep red, which I had mistakenly thought was blood. Very interesting, although I thought spitting this red fluid onto the sidewalk or street was not at all a pleasant sight.

§

It was on my first visit to Bassein (Pathein) in Burma (Myanmar) that the officers hosted another party, and this time, it was organized by the ship's captain, Laurie Cox. Our invited guests were the port doctor and his wife, our port agent and his wife, local dignitaries and residents, and relatives and friends of the invited officials. What made it spectacular was that our guests wore traditional clothing and the officers were all smartly turned out in the Red Sea Rig (black trousers and white shirt with epaulets) for this event. The boat deck was decorated with the usual colourful buntings, and the whole affair was very colourful.

Visible in the two group photos are Captain Laurie Cox, Chief Engineer Tony Lapsley, Second Engineer Reggie Xavier, Fourth Engineer Charlie Chan, First Officer Wally McNamarra, Second Officer John Stormont, Third Engineer Bob Tatz, Third Officer Jefferies and a young radio officer whose name escapes me.

Bassein (Pathein) is situated on the estuary of the Irrawaddy River. The river is fed from the Himalayas and is navigable for quite some distance from the Bay of Bengal. We left port fully laden with a cargo of rice. After we dropped off the pilot, our ship immediately ran into some choppy waters, and it felt like the ship had hit the bottom of the river. The bridge immediately rang down to stop engines, and Chippie, the ship's carpenter, went around taking soundings of each hold, looking for evidence of water leakage. Everything looked good and we then continued the voyage into the Bay of Bengal.

We docked in Bangkok for several days, and on a day off, Allan Noon and I took a bus to the River Kwai, visiting his uncle's military gravesite at Kanchanaburi War Graves Cemetery on the way. This was to be a full day's excursion.

We boarded a bus already bursting with happy, noisy people heading home to the countryside or visiting relatives there. They carried babies and had little children in tow, hauled baggage, baskets of vegetables and trussed-up chickens, and occupied every seat. We and many others had standing room only and had to be nimble to avoid colliding with our fellow passengers and the surrounding cargo.

The conductor wore a jungle-green uniform with a chest full of combat medals, a yellow lanyard on one shoulder, a Sam Brown belt (no weapon), flashy golden shoulder boards and a peaked hat with gold oak-leaf above the visor. He was incongruously barefoot below all this finery. I have no idea how he collected all the fares, because our bus was so congested that passengers in the back had no hope of reaching the exit door at the front, so instead, they had to use the rear windows to exit the bus, like we eventually had to do. It was all rather fun, and nobody seemed anxious or worried. To this day, I have no idea whether we even paid our bus fare.

Many young Australian soldiers are buried in the War Graves Commission Cemetery in Kanchanaburi; Allan's uncle was only 19 years old when he died. The bridge on the River Kwai (whose real name is the Mae Klong) was only a short distance from the cemetery.

There were two bridges at Site 277, one steel and one wood. We had to quickly get out of the way when a small locomotive came chugging towards us on the wooden one. We took some refreshment at a small teahouse at one end of the bridge.

Kanchanaburi in Thailand is home to the famous River Kwai Bridge. During WWII, Japan constructed the meter-gauge railway line from Ban Pong, Thailand, to Thanbyuzayat, Burma. The line passing through the scenic Three Pagodas Pass runs for 250 miles. This was known as the Death Railway.

The railway line was meant to transport cargo daily to India, to back up Japan's planned attack on India. The construction was done using POWs and Asian slave labourers in unfavourable conditions. The work started in October 1942 and was completed in a year. Due to the difficult terrain, thousands of labourers lost their lives. It is believed that one life was lost for each sleeper laid in the track. At the nearby Kanchanaburi War Cemetery, around 7,000 POWs whose lives were sacrificed in the railway construction are buried. Another 2,000 are laid to rest at the Chungkai Cemetery.

Allied Forces bombed the iron bridge in 1944. Three sections of River Kwai Bridge were destroyed. The present bridge has two of its central spans rebuilt. The original parts of the bridge are now displayed at the War Museum in Kanchanaburi.

Getting back to Bangkok, we made a short stop at the Wat Pho Temple to have a look at the famous Reclining Buddha. The temple was constructed in the reign of Rama III, emulating the Ayutthaya-style. The interior is decorated with panels of murals. It was quite a day, and we were glad to get back home on board our ship.

§

None of the officers went ashore on this visit to Madras (Chennai), which was my only visit to this area. It was here that I was the victim of a prank by my fellow officers.

Anchored not far from our position was a smart-looking Indian man-o-war. I was informed early in the day that the officers of this ship had extended an invitation to us to attend a reception in the late afternoon. Near the appointed time, I was all ready and dressed up in my full Red Sea uniform, but when I detected no action from my fellow officers (they were actually in hiding), I began to suspect a ruse, which it was! Later, over a beer, we all laughed at how gullible I was, but to me, this was all part of camaraderie, as it had no malice.

§

I visited Calcutta frequently over the next three and a half years, on three different company ships, but after my first couple of visits, I never had a strong desire to explore the city. India had been the jewel in Queen Victoria's crown, but it began to decline after the demise of the East India Company in the mid-nineteenth century. After India became independent in 1947, public property and buildings deteriorated further, litter was pervasive and poverty was rampant in certain areas. The Hooghly River, which runs through the city, is muddy and polluted; bloated animal carcasses are sometimes visible and rumour has it that human bodies have been sighted.

Company ships berthed at Kidderpore or Garden Reach, and sometimes the KGV Docks. As we tied up alongside our berth, hundreds of monkeys scrambled over the roofs of the dockside cargo sheds, shrieking and generally being a nuisance. Monkeys are considered sacred in India, as are snakes and white cows, so nobody shooed them off or paid them much attention.

Allan Noon and I explored the area near the docks one hot afternoon. The taxi driver dropped us off near Victoria Memorial, and we walked from there along Chowringhee Road, one of the main streets through the city. Before long, we were attracted by a commotion in the middle of the road. A white cow—the most sacred colour of cows—was lying down across a tram line, chewing her cud and gazing incuriously at the gathered crowd. Nobody could touch the beast, so traffic had to navigate around it. We didn't stay to see how the tram managed that.

The Victoria Memorial had, like many government buildings and monuments in Calcutta, been rather neglected since Independence in 1947. Its granite was grimy and weeds grew through the cracks in the stonework. It had been a grand structure in its day, but a lot of restoration work would be needed to make it right again.

Before returning to the ship, Allan suggested we stop for a cool drink in an air-conditioned cocktail bar, which I thought was a good idea. It was a quiet afternoon, and there was nobody inside the bar, so we enjoyed the undivided attention of the barman. Upon discovering that I was somewhat new to alcohol, the bartender made me a 'lady's drink'—a Pimm's #1 Cup, a smooth, gin-based cocktail served over chopped fruit, cucumber, mint and crushed ice. It tasted so smooth and refreshing, that it didn't take much for me to accept several refills before we departed for our ship. With my good judgment slightly impaired, I couldn't resist buying a full bottle of this cocktail mixture to take back to the ship, much to Allan's amusement.

In Calcutta, we acquired two passengers: an English couple travelling the world in a vintage London taxicab they had bought for £40. They had driven over land from England to Calcutta on the first leg of their circumnavigation of the world, and we were transporting them to Singapore, whence they would drive into China and up to Siberia. I believe their travels were sponsored by an international oil company and a tire company. We all enjoyed their company and were sorry to part with them in Singapore.

On another trip to Calcutta, Bill's wife accompanied him on the ship. Bill was the chief engineer on the ship at the time. Pat was a favourite of all the officers on board the ship, and she easily fit into all shipboard activities. I accompanied them to the Swimming Club in Penang and in Calcutta, and on other excursions during that trip. Bill and Pat also became my lifelong friends, and Bill would become the Best Man at my wedding in Hong Kong.

§

During my time on the Chun Sang, I experienced two machinery breakdowns that brought the ship to a stop mid-ocean. Both took place in the northeastern waters around Japan, and both times the sea conditions were not friendly.

In the first incident, the steering hydraulic telemotor on the bridge failed. The gear rack snapped in two, so the whole unit had to be dismantled. Our on-board facilities to deal with such a repair were limited: manual labour and basic tools such as hacksaw, rasp, portable electric drill and a set of thread dies (this was our most sophisticated tool) was all that was available.

While we worked, the emergency steering mechanism on the poop deck was engaged. Repairs took a couple of hours, and the Lloyd's Surveyor in Yokohama praised the engineers' repair job and suggested that the repaired gear rack be kept as a spare.

The second incident happened on a different voyage and caused us greater anxiety, because we had to shut down the main engine to conserve steam pressure in the boilers. On this ship, at sea, boiler feed water was delivered from a pump actuated by the main engine; an auxiliary independent feed water pump was used in port. First, the engine-driven pump failed, and when we engaged the auxiliary pump, it also failed. The engineers needed to make repairs without losing the steam pressure remaining in the boilers. Rough seas caused considerable difficulties when the ship was immobilized, but by concentrating on fixing the auxiliary pump, we managed to maintain steam in the boilers and produce enough power to give the ship sufficient steerage headway. The main engine-driven pump had to be safely disabled to allow the ship to continue on course until we arrived into Kobe.

§

The Chun Sang made regular calls to Kobe in Central Honshu, Japan, and we got to know the city quite well. There were many interesting shops along Motomachi Shopping Street, and there were hole-in-the-wall bars, karaoke singing clubs, English pubs and DJ dance clubs for those wanting a night out. Bars were numerous. A favourite was the CB Pub, owned by Courtney Brown, an ex-British serviceman, who told us there were 2,000 registered pubs in Kobe, and perhaps many more that were not registered.

§

A fellow called Sakamoto occasionally did repair work for us, and he always came by for a visit when we were in port and took a few engineers ashore for wining and dining. On one visit, the ship had arrived into port late in the afternoon, and as usual, the officers gathered together to have a few drinks after the engines were shut down. Sakamoto appeared with a helper, so we also poured liquor into both of them. It developed into quite a party, and by the time we decided to go out for dinner, it was about 9:00 p.m.

Sakamoto gladly offered to be our host for the night, and away we went. By the time we got to the first bar, I wasn't feeling too perky, but Sakamoto thought a little rest would fix me up—at a barber shop! I don't know how it happened, but later Sakamoto woke me up from my sleep on a couch. We then rejoined the rest of the gang from the ship in an interesting and exotic coffee bar, where the waitresses were all wearing bathing costumes (not bikinis). The coffee was not cheap when one considers that they were spiked! Before returning to the ship, we did eat dinner, and by then, I was ready to return to my cabin and into my comfortable bunk!

§

On another occasion, the ship did a trip to Hakodate on the northern island of Hokkaido. On the other side of the island, we also made a brief stop at Otaru, which is close to Sapporo. Bill Bennett and I took a little walk to watch some fishermen at work in Otaru.

Moji, located on Honshu (the southern island), was a frequent stop, mostly for cargo of cement. Women were very active in the process of loading our hold with bags of cement, walking in chain gang style, each carrying 112 cwt on their heads as they walked the plank from shore to the ship's holds.

In those days, Japan was always a favourite shopping destination for most of the ship's officers and crew, who eagerly loaded up with goods to bring back to their ship. Married men bought gifts for their wives and children, and sometimes shopped from a list supplied by the family, who were interested in china dinnerware, kitchen utensils, linen, towels, glassware, toys and clothing items such as kimonos and thong shoes.

I easily got caught up in this shopping fervour and bought plenty of stuff I had no practical use for—artworks, wooden cuckoo clocks, chinaware, battery-operated toys, transistor radios, glassware, lamps, tablecloths and sundry novelties. Having too many personal belongings could complicate transfers between ships, so I gave away most of my impulsive purchases to anyone who admired the stuff and received as much satisfaction from that as I had from buying it in the first place.

One of my purchases was a complete 12-place dinner set of beautiful Noritake chinaware, all carefully packaged in three or four cartons; why I bought it, I have no idea! Anticipating a transfer from the ship soon, I thought that I would store this with my sister. So, when next in Hong Kong, I shipped the packages by taxi to her, which had my name on them, but my mistake was not to include an accompanying note explaining my intentions. Months later, when I visited her, intending to retrieve my goods, she fell all over me with gratitude for my generous gift. What could I do? I just told her that I was pleased she liked it.

Shopping was thirsty work, and when it was done, we took our piles of parcels of all sizes to our favourite tavern, enjoying cool drinks while the proprietor and most of her staff examined with interest our purchases for the day. When it came time for dinner in the evening, off we would all go to look for a restaurant, leaving our purchases with the tavern staff and returning later to retrieve them.

We had a great relationship with some of the taverns and bars we frequented in the larger centres of Japan. There were no credit cards in those days, but if we happened to be short of cash, we were readily granted credit on the honour system. Sometimes, payment of a debt was deferred until our next visit, which might not be until a month or two later. Life was sweet and uncomplicated.

Because of my youthful naiveté, I was barred from late-night shenanigans with my shipmates. Instead, I was entrusted to collect everybody's parcels of shopping and return to the ship. I was never sure whether I ought to be thankful for their thoughtful care of me or miffed at being denied the adventures and life experiences they recounted later, back aboard ship. My turn came much later!

§

On June 2, 1953, Princess Elizabeth became Queen Elizabeth II, and we had a celebration on board the ship to commemorate the event. The party started in Singapore and continued all the way to Port Swettenham—about a 36-hour passage by sea. In honour of the occasion, the pennants flew proudly, and all the officers toasted the Queen energetically and repeatedly. Captain Kinnear, a teetotaller, was not party to all this frivolity, and kept himself secluded in his cabin instead, until the ship arrived safely into Port Swettenham. At one stage, the radio officer was seen prancing around the boat deck, talking to inanimate ventilator fixtures.

At the end of it all, Captain Kinnear was the only clear-headed officer on the ship; the rest of the officers must have surely paid dearly for their enthusiastic loyalty to royalty. The captain was very understanding though, and no official reprimand was ever received from head office.

Officers & Shipmates

Over the 18 months serving on *Chun Sang*, I met many officers, as the staff turn-over rate was high. Most of them were also advancing in their own careers. The list has been abbreviated.

Bill Edley was the first chief engineer I served under on board the *Chun Sang*. He was on the ship for one round trip, and because he was quite reserved, there was little opportunity to get to know him well. He was a married man with a wife residing in Hong Kong.

Bill Bennett, from Sunderland, England, was the chief engineer I served under the longest, and he turned out to be my greatest inspiration and closest mentor. I was always impressed with his willingness to don coveralls and joined his fellow engineers when we were working on maintenance in the engine room. He was one of the few engineers qualified with an Extra-First-Class MOT Certificate, which was equivalent to a BSc university degree in engineering.

Allan Noon, an Australian, was a second engineer, with whom I also served the longest. It was with him and Bill Bennett that we often made a trio on many shore excursions. Allan's wife Ruth and Bill's wife Pat were great friends in Hong Kong. Both these wives often met the ship upon arrival into port.

Laurie Cox, one of the captains during my time on Chun Sang, took a special interest in me personally. I remember having interesting conversations with him in his cabin from time to time. He was quite a gregarious person and I liked him.

Reggie Xavier was another one of my second engineers. He was originally a local Hong Kong boy who had served his apprenticeship at HWD. His mentor there was George Bolton. Reggie had just completed his apprenticeship while I started mine. I believe that Reggie was a volunteer with the HVDC and was made a prisoner by the Japanese in a military internment camp during the war. His wife Cynthia also lived in Hong Kong.

Tony Lapsley was chief engineer at the time, always smiling with an easy-going disposition. He enjoyed making new friends and participating at shipboard parties.

John Stormont was a second mate (second officer) on the ship, and he turned out to be a very special friend. John was a dour Scotsman from Dundee. Together with Allan Noon, Bill Bennett and sometimes Reggie, we often did things together on shore at some of the ports. John eventually became master of his own ship, and before retirement, he was appointed as a Director of Jardines, Matheson & Co. in the London Office.

Jack Perrin was a mate (chief officer), originally from the UK. I found him to be exceedingly jovial, and we got along well together.

Wally MacLauchlan, from Newcastle in England, took over from Jack as mate on the ship. Wally used to enjoy bantering me from time to time, and I found him to be an excellent shipmate.

'Batchy' West was a radio officer on the ship, and he and I got along very well together. I had a great respect for him not only because of his age, but because he was such a gentleman, and to some extent, a confidant for me. We played a lot of chess together, and he taught me cribbage; I also admired him for his agility at deck tennis. I had enormous respect for him.

Sailing with such an agreeable complement of ship's officers was one of the highlights of my career, if not my life. My experiences were similar on other ships, and with the other crew members, who were mostly Chinese. The engine room crew were mostly big guys from Tsingtao in Northern China, and the stewards and kitchen staff were mostly Cantonese, from the south.

My lengthy experience on the Chun Sang opened the world to me, a world that had been denied to me through all my growing-up years. What a wonderful experience!

Farewell to *Chun Sang*

I felt sad at leaving the Chun Sang, my home away from home for the longest time on any ship I would ever serve in my sea-going career. The fellow officers were the greatest, and amongst them would be several lifelong friends. How appropriately that 'spirit growing' is the translation for Chun sang.

On October 22, 1953, I was granted leave to prepare for the final examination for the Second-Class Certificate. I took my leave in Hong Kong, and stayed at the European YMCA in Tsim Sha Tsui, where I made my final preparations to sit for the examination.

The examination took place at the Government Marine Offices in Hong Kong. Mr. Buckley presided at the examination, and the only thing I can recall was he had one glass eye, and I was never sure where or who he was looking at during the oral examination. I successfully passed my final examinations at age 22 and reported this success to John Green at head office. This meant that I was now officially deemed technically competent to be the second engineer on any British merchantman of any tonnage or horsepower rating. I was also accepted as a graduate engineer of the Institute of Marine Engineers in London.

SS Eastern Queen: The Company Flagship, November 1953 – July 1954

SS *Eastern Queen*, the flagship of the ICSN Co. Ltd., pride of Jardine, Matheson & Co. Principal particulars: Built by Denny's Shipbuilders in Dumbarton; Year built 1950; Gross Tonnage 8,644 tons; 8400 HP; Single screw driven by PAMETRADA Compound Steam Turbines; Two B & W WT Boilers rated for 650 psi and Superheat to 650° F. This superb snapshot was taken in Saigon in 1954 when the ship was fully loaded and preparing to sail for Hong Kong.

My assignment to this ship was not only a shift from the smallest ship to the largest ship in the fleet, but it was also the most powerful and newest ship at the time. It was built in 1950 by Denny's Shipbuilders in Dumbarton, Scotland, with a gross weight of 8,644 tons and powered by a PAMETRADA compound steam turbine rated at 8,400hp. Two high pressure water tube boilers supplied steam to the turbines.

The ship was classified as a cargo liner and provided 24 staterooms for first-class passengers, and accommodation for an additional 28 second-class passengers. She also had space for about 500 steerage passengers in the 'tween decks. The trade route was Calcutta-Straits Settlements - Hong Kong - Japan.

I was awestruck with everything about the ship when I first came on board on November 21, 1953. Compared to the two previous ships, it was like moving from a one-star hotel to a five-star hotel—it was dazzling! The machinery space had the same effect on me.

Of course, the equipment was all new to me, and I found it to be most impressive. As on the Chun Sang, but to a greater extent, manoeuvring the ship in and out of port required two engineers, and sometimes even the chief engineer was in attendance in the engine room. I was happy that I was appointed sixth engineer, which gave me time to learn the engineering intricacies in the ship, with the lowest level of responsibility.

I reported to Jimmy Lindsay, who was the chief engineer. Then I met Chuck Eikoff, the second engineer, who was an Australian from Adelaide. I think he must have been new to the company, as I had never heard of him before. Apart from the chief engineer, there were only two other engineers with a more senior certificate than mine, and that was Chuck Eikoff and Eric McGarva; I had never heard of Eric McGarva before, and he might have been a new hire from the UK. In total, there were six watch-keeping engineers, and sometimes we would have a couple of spare engineers as well.

It was when I was serving on this ship that I had made the purchase of a cine camera (another one of my frivolous purchases). I purchased a Eumig 8mm Cine Camera from Filmo Depot on Queen's Road. Henry and Poldi Corra were the owners and operators of this business. I first met their daughter Christine at Mrs. Rodger's party on Coombe Road many years before, and in Stanley Camp.

Calcutta Bound

As the *Eastern Queen* was on the same trade route as the *Chun Sang*, I would be visiting some of the same ports, but this time in the company of a new group of officers.

Under the command of Captain Schofield and First Officer Wesley Bartlett, better known as 'Bart', the ship sailed with passengers for various destinations on this route. I remember there was a missionary family, Calcutta-bound, and an elderly lady under medical care who appeared to be of the Ruttonjee family in Hong Kong; she had some younger family members in attendance. I had knowledge of the Ruttonjee family from the war when I was a refugee at the Canossian Refugee Centre on Caine Road. The Ruttonjee family used to be generous benefactors to various charities in the colony, including the Italian convent. I remember seeing Dhun Ruttonjee from time to time; they were of Parsee origin.

Having adjusted to the new machinery, I began to get acquainted with my fellow engineers. Chuck Eikoff, the second engineer, initially seemed very pre-occupied with his responsibilities and was quite serious most of the time. Jimmy Lindsay, the chief engineer who I had already met, was quite senior in the company and not far to go to retirement. I found him relaxed and easy to get along with. The one that impressed me the most was Jack Pettigrew, who had joined the ship in Scotland when it was completing construction.

Jack Pettigrew was the third engineer, originally from Glasgow in Scotland, who had been with Jardines for several years. He was called a 'permanent third' because he was never likely to pass any exams. His value to the ship was his profound engineering knowledge with the machinery on the ship, and this would very soon stand me in good stead. But first, I would have to understand his difficult nature, which was augmented by his propensity to imbibe Scotch whiskey.

I seem to recall there were frequent gatherings in Chuck's cabin, many of them of a social nature. Then there was Eric McGarva, a very sociable person, who I think was the supernumerary second engineer. I remember well that at some of the social gatherings in Chuck's cabin, there was not enough sitting room for everybody, and because of this, I often found myself perched on his bunk. I mention this because it is an important point later.

Upon arrival into Calcutta, the passengers reaching their destination readily disembarked after clearing through customs and immigration. The engineers secured all machinery in the engine room for port operation, while the deck officers were busy handling documentation with visiting officials.

It is customary in Calcutta to witness customs officials searching the ship whenever they felt like, and it all seemed rather random to me. But I had no comparisons from the past to make this judgment. What was about to happen would turn out to be quite dramatic.

The customs officers at this time were particularly active about our ship and seemed to be expecting to discover some significant illegal contraband. Sure enough, it was in the second engineer's stateroom that they found what they were looking for. Hidden on the floor underneath the built-in bunk they found a stack of solid gold bars (about 200 'fingers'). The second and chief engineers were immediately detained at the Custom House on shore; the chief engineer was eventually released, but the second engineer was arrested, and we never saw him again.

Then the customs officials started interrogating the entire group of engineers, one at a time, perhaps feeling that there might have been some collusion with the second engineer. When my turn came for 'undivided interrogation' at Custom House, I learned the reason why I might have been a suspect before the fact. Their interrogation seemed to be centred around the knowledge that I had often been sitting on

Me, and two fellow officers having fun. Left is George Taylor, 3rd mate, and right is Eric McGarva, 2nd Engineer. c1954

Me peering out of my stateroom window. This was the flagship of the fleet, and I felt privileged to experience my first senior rank on this ship. c1954

Nightlife with the "boys" in Kowloon. Joined with officers from another of the company ships in port at the time. Next to me on the left is John Sullivan, and third from the right is Rene Ruyters. Here again, I had no experience in night life in Hong Kong due to my sheltered upbringing under Godmother. c1954.

Eastern Queen, ship's officer partying in the Officer's Lounge somewhere in Japan 1954. Back row: Reg Smith, Radio Officer; Jack Pettigrew, 3rd Engineer; Jock McWatty, Engineer; Eric Norman, Mate; Me, Acting 2nd Engineer. The huddling trio: Paddy Thompson, 2nd Mate; Eddy Oliviera, Engineer; "Wee" Jock Simpson, 3rd Mate.

the second engineer's bunk, and perhaps I might have had knowledge about this contraband shipment After severe interrogations, which lasted from mid-afternoon to two or three o'clock the next morning, I refused to cooperate anymore and demanded that our company agent come to Custom House to affect my release from any further interrogations.

With two senior engineers removed from the ship, Captain Schofield, the master of the ship, was faced with a decision. The choices available on the ship at this stage were now one qualified senior engineer with experience (Eric McGarva), one qualified junior engineer with minimum experience (Bob Tatz), and one 'junior' engineer with solid experience, but no qualifications (Jack Pettigrew).

John Green in head office was busy trying to cover all the bases by shifting engineers from other ships in the fleet. In the meantime, for the next seven months, I found myself filling the gap on the *Eastern Queen* as follows: 20 days as sixth engineer; 3 days as fifth engineer; 14 days as acting second engineer; 27 days as fifth engineer; 39 days as acting second engineer; 21 days as junior third engineer; 69 days as third engineer; and 8 days as second engineer. I gained a lot of experience during this period, and my most valuable support came from

Jack Pettigrew, even though he was breathing whiskey fumes in my face when I was seeking his cooperation from time to time. Fortunately, the ship did not endure any unexpected mechanical failures during this entire period.

A magnificent surprise Christmas present was going to be sprung on Bart, the chief officer. Unknown to him, Mavis, his wife, flew into Calcutta from Hong Kong, and with the help of the ship's officers, she was smuggled on board the ship. That evening, he was surprised by her entry into the dining room for dinner. The look on his face was priceless, and everybody joined in the happiness flowing all around us. The celebrations were very festive. That weekend, they both flew away to Darjeeling in Nepal to spend some quiet time together. In the meantime, the ship's officers went shopping in the Grand Bazaar, and brought back a rocking horse to be presented to their daughter Caroline when the ship returned to Hong Kong.

The Gurkha Regiment

To combat the communist insurgency in Malaya between 1948 and 1962, Britain had established a security force to conduct offensive operations against communist terrorists. Security forces consisted of police, military and commonwealth troops. It was in connection with this objective that a Gurkha regiment, together with their wives and families, were being transported on the *Eastern Queen* to Singapore, from where they would be deployed somewhere in Malaya.

The Gurkha regiments originated in Nepal and are British-trained; the embarkation port was Calcutta. Before disembarking in Singapore, the commanding officer of the regiment presented a kukri knife, suitably engraved, in a sheath to Captain Schofield in appreciation for the service provided to the Gurkha officers and their families. We had the pleasure of interacting with some of the family members, and the women folk were beautifully dressed in their traditional clothing. We also had the opportunity of interacting with some of the children, but unfortunately, there was very little dialogue because of the language barrier. I was able to record a short 8mm movie.

Singapore – Saigon – Japan– Hong Kong

On this voyage heading for Hong Kong, we also made a stop at Saigon. At the time, Indo-China was still under control of France as a colony. One early afternoon, I took a walk down Rue Catenat, the main street in town, and was quite amazed at the architecture and wide avenues stretched before me. I also noted that the streets were practically devoid of any humanity and traffic, not knowing that the French here practised an extensive midday siesta. I was lucky to find a wine cellar open (known as caves). With nothing much else to do, I bought a case of Chablis wine and brought it back to the ship.

I don't know what our ship's cargo was (perhaps rice), but before the ship set sail for Hong Kong, I took a series of beautiful photographs of the ship from a rowboat taxi on the river. The ship was loaded right down to her load-lines!

I think the Chinese New Year in Japan was pretty much the last fling I enjoyed on the *Eastern Queen*. It was either in Nagoya or Kobe that we had discharged cargo, which allowed us several days in port. I was the acting second engineer at the time.

A Mr. Chan was the comprador on the ship, and he had his own quarters on a lower deck. I understood that he was a very important man, as movement of all cargo on the ship was processed through him. For the Chinese New Year celebration in 1954, he had put on a huge Chinese dinner for all the officers. By the same token, the Chinese bosun and the engine room head fireman also put on a feast for the officers. What with the Northern Chinese food and the Cantonese food, we really felt stuffed before it was all over; the copious supply of liquor helped to wash down all this food.

Hong Kong Social Visits with Godmother

On one of my visits to my godmother, I joined her on a visit to Babushka in the Matilda Hospital, where she was under care for a broken hip. The hospital surroundings were very familiar and brought back emotions and poignant memories from my experiences there at age seven. Babushka was old and frail and did not heal well, and never recovered; she eventually died in the hospital.

On another visit to Hong Kong, I joined Godmother and Harry on a visit to Dorothy Tipple and her daughter Lesley. Dorothy drove us on a day excursion in Kowloon in the New Territories. We made several stops to admire the countryside before stopping at the Shatin Roadhouse for refreshments. My association with the Tipples, begun before the war, continued well into adulthood.

I well remember one evening I had invited my godmother and Harry to have dinner with me at Gaddi's restaurant in the Peninsula Hotel. My sister and her husband were also dining there at the same time, and when they saw us, they kindly sent a waiter over to our table to offer us a complimentary drink. They were a little taken aback and somewhat amused when Godmother brazenly ordered a full bottle of a good table wine instead of the customary single glass of wine.

Officers and Shipmates on the Eastern Queen

Captain Schofield: I would be serving under him the longest during my career with Jardines, on several different ships.

Michael Pope, a second mate, who later met a tragic death in Hong Kong.

Ron Learoyd, also a second mate, who I would sail with again on two ships.

Wesley 'Bart' Bartlett, the first mate (chief officer) and later when he was captain on another ship.

Paddy (George) Thomson, second mate, a wild Irishman but with a warm disposition. He loved his Guinness, was always ready for a party and never minced his words, saying what was on his mind. There is a story about Paddy when I was visiting Dublin during my first long leave.

Jack Pettigrew was considered a 'permanent' third engineer, but very knowledgeable and experienced. He had been assigned as part of the team taking delivery of more than one new ship from the ship builders in his career with Jardines.

Eddie Oliviera and his younger brother Tony, between them, would serve with me as engineers during my period on the ship; all three of us had served our apprenticeship together in HWD.

SS Hew Sang 曉 生 *(Dawn Growing)*: July 20, 1954 – November 14, 1954

I was going to be on the *Hew Sang* for just under four months. The trade route was new to me. The main business was shipping logs from British North Borneo to Japanese ports; from Japan to Hong Kong. Some general cargo was carried, and I can only remember the ship loading cement in 100 kg bags in Moji; these bags were carried by a female chain gang from shore into the ship's hold. Then from Hong Kong back to British North Borneo with general cargo, usually building materials.

The company owned and operated a total of five similar ships, which became affectionately known as the 'H' Boats. Hew Sang was one of them. They were built by William Gray Yard in West Hartlepool and engineered by Central Marine & Engineering Works (CMEW) of West Hartlepool. They were acquired by Jardines between 1946 and 1955; Hew Sang was acquired in 1949.

With a gross tonnage of 3,539 and 1,500hp, it was a far cry to what I was used to on the *Eastern Queen* and took me all the way back to the Wing Sang, my very first ship. Another similarity to the Wing Sang was in the person of Paddy Richardson, my first chief engineer, except that under him now, I was the second engineer. The propulsion machinery was very simple—triple expansion uniflow reciprocating steam engine, with steam supplied by two oil-fired Scotch marine steam boilers.

I think I did two or three trips on the *Hew Sang*. The ship had an unusual superstructure design, which made it look like a small oil tanker or lighter boat; it looked quite ungainly. Officers' accommodation was located mid-ships and the engine room was at the rear of the ship. To reach the engine room to go on watch meant that the engineers had to walk the open deck. None of that mattered to me because it was going to be a new experience, and that's what I wanted. Furthermore, the ship was on the timber trade from Borneo to Japan, and this was to be my first visit to Borneo.

In those years, the island of Borneo consisted of large tracts of undeveloped territory, which was heavily forested with exotic timbers. Because there were no roads connecting major communities, population growth was inevitably centred along the coastline, where reliance was placed on communication by boats. Within principal communities, there was a system of paved roads, but it never extended much beyond the perimeters of the towns or villages. By this time, air transport was gaining popularity, following the development of airports in principal centres.

All ships visiting these ports were warmly welcomed, and although we were not the bearer of supplies to the communities, our reception was no exception. I still have fond memories of several memorable experiences with the local inhabitants.

The principal ports that I remember in Borneo for this period were Sandakan, Labuan and Jesselton, but there might have been others. I know that we used to refuel in Tawau. I also became familiar with the name of an American well-known in the business, Don Ireton, who had a residence on an island bearing his name. He was from Oregon and knew the timber business well. Also, he came through this part of the world as a senior officer under General MacArthur on the march towards Japan in the Second World War.

For a change in pace, ships' officers used to visit the bar in the Sandakan Hotel. There was a nice swimming pool and the service was quite good. Sandakan had a regional airport, which was serviced by Malaysia Airlines out of Singapore. I was totally surprised when, one day, amongst the air crew checking into the hotel, I saw Iris Jantzen. I immediately reconnected with her, and we had a joyful reunion talking about the old days, and the parties on the Chun Sang in Singapore. This would be the last time that I would have any connection with Iris, but the memories are good.

My discharge from the *Hew Sang* took place in the Port of Kobe on July 20, 1954. From there, I went by train to Kure, via Hiroshima, to sign onto the Wo Sang. However, I would be returning to British North Borneo after my return from the first long leave, and curiously, it would be on the Hew Sang again.

SS Wo Sang 和 生 (Harmony Growing): November 1954 – September 1955

The ship was on contract to the UK Ministry of War Transport (MOWT), together with the E Sang (怡 生). The service was to exclusively provide transportation to meet British Commonwealth requirements as needed in the Korean theatre under the auspices of the United Nations.

Wo Sang was a pre-war ship built in Glasgow in 1934 with a gross tonnage of 3,539. Main propulsion machinery was a triple expansion uniflow reciprocating steam engine rated for about 3,000hp. Service was on charter to the Ministry of War Transport (MOWT).

There was first-class cabin accommodation for military officers, and 'tween deck accommodation for regular troops during a sea passage of about 15 hours. The chief engineer was Norman McQueen, Frank Lee was the third engineer. I don't recall the name of the fourth engineer, who was Chinese.

The Korean War was in a state of truce at the time of my service, and except for considerable troop movements, there was very little exposure to hostile action. The US Army had a significant presence there, and we soon got to know some of the officers of the 501st Transportation Unit located in Pusan.

The transportation of troops and equipment involved two voyages a week between Kure on the Japanese Inland Sea to Pusan on the south coast of Korea, an approximate distance of about 265 miles. On one occasion, we also went to Inchon. We also transported British military war dead in coffins to Kure. The loading of flag-draped coffins into the hold was always an impressive ceremony of respect paid to the fallen soldiers, accompanied by the mournful sound of bugles rendered by honour guards. The company had also provided a second ship on this contract, the *E Sang*.

Ship's officers frequented the US Army bases in Korea and benefited from their post-exchange (PX) stores. In return, US Army officers always accepted an invitation to dine on board our ship. The white linen tablecloths, sterling flatware, fine China crockery and being waited on by liveried stewards was a welcome change, together with the ready supply of all kinds of liquor as a bonus. I believe that access to liquor in Korea was very limited for US Army personnel. The ship's officers also entertained the ladies of the British Red Cross and the Canadian

At historic 38th Parallel in South Korea. The Military Demarcation Line (MDL), sometimes referred to as the Armistice Line, is the land border or demarcation line between North Korea and South Korea. On either side of the line is the Korean Demilitarized Zone (DMZ). The MDL and DMZ were established by the Armistice at the end of the Korean War in 1953. This condition continues to this day as no permanent peace accord has been signed. I am on the right.

Wo Sang taken in Kure, Japan with the female shipping agent, and some of the ship's officers. I am next to the agent on her left.

Women's Voluntary Service who were based in Korea and Japan; this included British army officers.

In Inchon, Captain Cy Feres from the Corps of Royal Canadian Electrical and Mechanical Engineers was kind enough to invite several of us to a tour of Gloucester Valley at the Imjin River Bridge Crossing, on our way to the 38th Parallel marker. All the terrain that we saw was rather bleak, and we learned that the Gloucester regiment earned seven VCs in a battle against Chinese troops in 1951.

We stopped briefly at the 38th Parallel and saw no buildings or any sign of military presence. The British had a base camp about 20 kilometres away, where Cy Feres made a brief stop before returning to Inchon via Seoul.

Bath House Experience in Kure

My bathhouse experience in Kure only occurred because of a problem with our single fresh water pump. It was a steam-driven pump with a very unusual control valve system and had broken down but stubbornly defied all resolution, no matter how hard I tried to solve the problem. When I put down my tools late that night, I thought I would investigate—for the first time—the pleasures of a typical Japanese bathhouse, with the idea of getting cleaned up and returning to the ship immediately afterwards.

The quartermaster at the gangway hailed a taxi for me, and getting in, I told the driver to take me to the best Japanese bathhouse in town. Because it was dark outside, I had no idea what route he was going. I was surprised when he pulled up at what looked like a regular Japanese bar adorned with glittering neon signage, especially when I could see American sailors coming out of the door. Only when the hostess convinced me that they did offer bathing facilities at this establishment did I let the taxi go.

At the reception desk, the hostess asked me whether I would like the full treatment. Without knowing what that entailed, I agreed, wanting so badly to get cleaned up quickly and get back to my ship. Pushing my way past several sailors crowding the desk, I followed my hostess into an inner sanctum at the back of the establishment, and there I was handed over to another attendant dressed in a kimono.

I was shown a cubicle, not unlike a sauna, and urged to undress for my wash. I had never undressed before a lady in my whole life, so I waited for her to step out of my space before starting to shed my clothing. When she returned with some towels, I stopped undressing and was aghast when she stood by, waiting for me to continue. Almost at the same time, another attendant walked in with more materials (soap, etc.), so the situation developed into a little bit of a stand-off. The two young ladies were giggling between themselves as they were waiting for me to undress, while I was concerned about my nakedness. Thankfully, they understood my bashfulness and left me to wash myself. When they came back, they did help me to finish drying myself.

I refused their offer of additional treatment, even though I had paid for it. Instead, I wondered, by sign language, if they could provide something to eat. Completely dressed now, I was led into a room opening out onto a courtyard; in the courtyard was a beautiful little garden with a small pond in the middle. A full moon was shining down on this scene, which was also lit up by twinkling electric lights.

I was invited to sit down on the floor in front of a low table. Before long, an elderly lady, followed by two women attendants also dressed in kimonos, came into the room and sat down beside me. This was going to be entertainment while the food was being prepared. They played music, sang Japanese song and offered me heated Japanese rice wine. I thought to myself, "I better have enough money to pay for all this, if I run out of credit for the full treatment."

Food came: sushi and a hot bowl of meat and vegetables with rice and other ingredients I did not recognize, and enough to also feed the three or four ladies giving me their undivided attention. I thought this was great and started to enjoy their company, by sign language on my part, and smiles and giggles from the younger ladies.

At the end of it all, there was no extra charge, because there was sufficient surplus credit from the unused portion of my entrance charge. When my taxi arrived, all the ladies gave me an enthusiastic send-off and saw me to the doorway. I didn't get back to the ship until about two in the morning, feeling tired, but clean and well-fed. The next morning, none of my shipmates would believe my story, and furthermore, I could never locate that establishment again.

§

The ship entertained most lavishly during the Christmas season and other traditional public holidays, with members of the armed forces in the area, as well as the military nursing staff. It was inevitable that intimacies led to commitments, and two of our officers, one on the E Sang, and the chief officer on our ship, John Adair, would end up married. On these occasions, our guests would also reciprocate with their own generosity. I remember attending Christmas dinner in one of the American camps along with my friend, Paddy Behan, the ship's radio officer.

A British army captain in Kure had organized a launch party on the Inland Sea of Japan, which included some of the female service personnel. We didn't do any swimming, but we did drink a lot, together with picnicking on a beach. Lots of fun was had by all.

It had been only three and half years since I had broken away from a very confined existence under Godmother's control. Up to now, my character development faced numerous challenges, and understanding women was only one of many.

An important priority for me was to prepare for engineering exams to enhance career advancement with the company. Until this was fully achieved, I did not allow myself to be distracted. There had been no room for any romantic relationship development at this stage of my life.

By now, I had many opportunities to witness interactions between my fellow officers and their female companions in various situations, be it in a sedate social setting or a wild night on the town. The actions shown by married officers versus their unattached colleagues were markedly different; I belonged to the second category, but because of my priorities, I behaved like those in the first category.

Paddy Behan, the radio officer on my ship, also belonged to the second category, but because of his commitment to a betrothal, he belonged to the first category as well. We were good friends and shared a lot of common interests. He often discussed the attachment he had with his girlfriend in Ireland, including his vision of the benefits of marriage with her. Anybody older than me was like a god, so I listened attentively, mostly out of politeness.

Introduction to London Pen Pal – Eilish Hannigan

When Paddy learned about my plans to go to London for my first long leave, he asked if I would be interested in meeting a girl. Apparently, Paddy's Irish girlfriend had a close friend who had recently terminated a relationship and was now unattached and working in London.

Me and Paddy Behan in Kure, getting ready to go ashore. Paddy was the one who introduced me to my future wife, and it was here that we started our penmanship. c1955

First introduction by this photograph to my Irish pen pal, who subsequently became my wife.

The idea appealed to me because my contacts in England at the time were very minimal: there was my sister Margaret, recently widowed and living in Birmingham, and my friends Reg and Marjory Kirby from my first visit to London in 1945, now living in Saffron Walden.

I discussed this with Paddy and told him that my main preoccupation in London was to attend an engineering course to prepare for a very important examination; nothing must interfere with this. I presume he conveyed this information to his girlfriend in Ireland.

A positive response eventually came back through Paddy, signifying interest in initiating preliminary correspondence, and that was how Eilish and I started our penmanship relationship. For the next few months, Paddy and I had a common interest to share with each other, until he himself went home to Ireland on leave.

The intermittent correspondence between Eilish and I was for a period of about six months, all via 'snail mail' through the shipping agent located in the Kure Office in Japan. The correspondence, as I recall, was mostly about my work and career aspirations. There was very little to discuss about my family and heritage, as at that time I myself was still largely ignorant on this topic; it was many years later when research grudgingly revealed additional information. That didn't bother me at the time, because my interest was in the future, and what kind of life I visualized for myself. Eilish didn't reveal too much about herself, and that was the way it had to be under the circumstances. The only intimacy was an exchange of photographs so that we would recognize each other later. Eventually, I found out she was 4 months older (June 18, 1931).

Departing *Wo Sang* in Preparation for First Long Leave

The day came when I was relieved from the *Wo Sang*. I had spent 10 months with a fine group of fellow officers. I would particularly remember Paddy Behan and Norman McQueen, whom I had the pleasure to become friends with. Unfortunately, I would not see either officer again, as our paths were not destined to cross each others'.

On September 14, 1955, I left Hiroshima and travelled by train to Kobe, to join the *Eastern Queen*, and signed on as supernumerary second engineer on September 15, 1955, arriving in Hong Kong on September 30. I wasn't required to stand a watch, as the engine room was fully staffed at the time. The chief engineer was Bill Beasley.

In Hong Kong, John Green placed me on reserve status from October 1 to October 20, 1955. This gave me time to visit my godmother and Uncle Harry on Prince Edward Road; by then, Babushka had died in the Matilda Hospital.

Godmother and Remnant of Family Inheritance

Totally unexpectedly, on my last visit to Godmother and Uncle Harry at Prince Edward Road, I was presented with certain family assets that I had thought were irretrievably lost at the time of Mother's death. I had no idea how and when Godmother had acquired possession of these items, and at the time, I did not think of asking any questions.

As far as I can remember, this treasure trove consisted of a collection of Mother's artwork on canvas and parchment, and charcoal sketches; two or three rather worn family photograph albums and numerous loose photographs; certain documents in Russian, probably pertaining to my mother; a package of what might have been documents, tightly wrapped in a waterproof package; a copy of Mother's British passport and naturalisation documents; my First Communion certificate; my birth certificate; and there may have been more.

I remembered some of the paintings, especially the portrait of the blind beggar she rescued from the street when we were living on Nathan Road. The photograph albums were also familiar because, as children, my sisters and I used to pore over every page, squealing with delight when we were told which one was of us.

Still surprised by this revelation, I listened to Godmother say that it was now about time that I had custody of this material. No doubt, Godmother did not want to be burdened with caring for all this material, as they were planning to soon leave Hong Kong to retire in Australia.

"After all," she continued, "you are the only male survivor in your family, and you are fully entitled to all these assets. Under no circumstances are you to share any of it with your sisters, Julie or Margaret."

I did not express my concern with Godmother at the time, but there was no possibility for me to store all this material on board any of my ships. There was no storage space in my accommodation, and anyway, it would have been too cumbersome to move them from ship to ship. Despite my godmother's earnest adjudication, I had no choice but to entrust all this material to my sister Julie, with the understanding it was to be only for safekeeping.

I chose to keep all my personal documentation, the family photograph albums and the collection of loose photographs; this I could manage with my personal effects on board my ships. My relationship with my sister at the time was quite cordial, so, leaving Godmother, I took a taxi directly to Julie's home at Nairn House on Waterloo Road to drop the bulk of the material in her care. She too was astounded and gladly accepted the lot. I made it clear to Julie at the time that all this material had been given to me by my godmother, and that I would want them back at some future date.

Many years later, when I tried to claim some of this material from her, Julie refuted my claim, although she could never explain how these family assets came to be in her possession. My mistake at the time was that I did not return shortly afterwards to document a complete inventory of all this material in her custody. Without proof to support my claim, I would almost certainly come out the loser in any confrontation or dispute, so I backed down.

Summary of My First Jardine Contract

Yes, I did receive regular bank statements, but apart from cursory glances, I never worried about the balances. All I knew was that I had no debt, had a bank account in the UK, where my salary for over three and half years had been accumulating, except for the monthly remittance to my godmother. The prospects of six months' leave with full pay was exciting; I was ready for whatever lay ahead. I might even visit Margaret, who was recently widowed when her husband Jimmy died from an aneurism in the brain. My sister was now a single mum, looking after her son David, who was about five years old.

The last three and a half years had been full of lessons, both professional and personal. I had come of age; I had established an enviable reputation with Jardines. I felt fulfilled and ready for the next step. Surely there were many more lessons to learn—I still had a long way to go to fully understand what maturity meant—but I was satisfied with the direction my life was going at the time, especially compared to my life before Jardines.

CHAPTER 12: LIFE CHANGING EVENTS

Long leave in the UK

October 1955 – May 1956

My first long leave began on October 21, 1955. The voyage counted towards my six months of leave, so I had five months in England to accomplish what I had in mind.

I sailed on the P&O liner *Carthage* from Hong Kong and arrived in London at Tilbury Docks on November 12, 1955. I was giving a lot of thought to what I would do in England. There was my sister Margaret living in Oldbury, Marjory and Reg Kirby living in Saffron Waldon and attendance at my course at Poplar Technical College, followed by examinations; what if I had to repeat a failed examination? I had no idea how my relationship was going to go with Eilish Hannigan—an unknown factor!

Tony Lapsley, formerly my chief engineer on the Chun Sang, was also going on leave with his wife to the UK on the *Carthage*. He was travelling first-class because of his seniority; I was in tourist-class. However, he and his wife preferred the entertainment and the less stuffy atmosphere of the tourist section, and that gave us the opportunity to visit a few times.

The passengers on the ship I associated with were a group of Hong Kong policemen, mostly bachelors, also going on leave to the UK. They were a congenial crowd, and at times, we did shore visits together, particularly in Colombo, Bombay, Aden and Gibraltar.

In Colombo, I visited the Botanical Gardens with a couple from the ship, and impulsively bought a nice silver necklace complete with a single moonstone pendant. At the back of my mind, I could see myself giving this to Eilish, no matter how our relationship would move. As it turned out, it worked propitiously.

In Bombay, I joined the same couple for a mini tour of the city. There we watched a snake charmer perform with a cobra and a mongoose. He didn't allow them to fight to the death, but it was a gripping conflict. Then we also visited a Parsi Tower of Silence on Malabar Hill, located in a posh residential district. A Tower of Silence is a circular, raised structure built by Zoroastrians for excarnation—for carrion birds to consume the flesh of the dead. The corpse is laid on a metal grille at the top of the tower, and the bones fall to the bottom when the vultures have picked them clean.

The stop in Aden was brief, and it was my first introduction to camels and real-life Arabs. Our stop in Gibraltar was also very brief, and my main impression was of the vast numbers of free-roaming monkeys, reminiscent of my first experience in Calcutta in 1952.

On November 12, the *Carthage* tied up at her moorings at Tilbury Dock, where all passengers were processed by immigration and customs. Tony and I shared a taxi to take us to our hotel in Lancaster Gate, in the heart of Central London. After registering at the Merchant Navy hotel, I think we took some supper and stayed overnight.

As Tony and his wife were on a different agenda to mine, the next day they commenced their holiday tour around the British Isles. As for me, I made my way to Eastham to locate Poplar Technical College.

A Mr. Sinclair and a Mr. Hodge were two instructors that I remember well, lecturing in the Marine Engineering Department to prepare candidates for MOT examinations. I was already familiar with most of the curriculum for the course I was enrolling in, as I had been preparing myself beforehand by studying on the ships.

The college also directed me to suitable student accommodations in the nearby area, and I soon became acquainted with a kindly lady named Mrs. McGee. She offered students breakfast and evening tea with the price of the room. For lunch, the Ann Boleyn Pub—conveniently located near to Mrs. McGee's place—was the most popular venue. The pub also had a magnificent hearth by an enormous fireplace located in the public area. I checked out of the Merchant Navy hotel and moved my gear to Mrs. McGee's room.

Meeting Eilish Hannigan

Eilish was in residence at a ladies' hostel in Spanish Place, not far from Madame Tussaud's Museum on Baker Street. The residence was run by a religious order for single women working in London. She was admittedly nervous at this our first meeting, held in the evening in the parlour of the residence. We talked about Paddy Behan and Mary Brophy, and a little bit about Hong Kong; it wasn't long before she warmed up, especially when I presented to her a gift I had purchased in Hong Kong. It was a Chinese fleece-lined turquoise-coloured silk brocade dress jacket. She appeared delighted with this gift, but very demurely did not try it on before me!

Understanding that I was busy studying for an examination, we parted amicably with a promise that I would be in touch before long to arrange a get-together, maybe on a weekend. On the way back to my digs in East Ham, I couldn't help thinking about this meeting. I was impressed with her composure, friendliness and her beautiful smile, and I decided to myself that I wanted to get to know her better.

With Eilish working and me in classes during the week, we didn't see much of each other in the first few weeks. However, an opportunity later presented itself for me to arrange a date. With her agreement, I purchased tickets for the two of us to attend a Royal Command Performance by the Bertram Mills Circus, on Wednesday, December 21. Eilish was a great admirer of the Royal Family and was delighted to be seated not far away from the Royal Enclosure, where she could easily view Queen Elizabeth II and Prince Philip. As a bonus to me, Eilish looked stunning, wearing the brocade jacket that I had given her at our first meeting.

After the performance, we had supper at Lyon's Corner House in Piccadilly Circus. We talked about the performance and how nice Queen Elizabeth looked, and the chat was going very well when, with a sudden start, Eilish suddenly reminded me that she must get to the hostel soon because of the 11:00 p.m. curfew. Well, we didn't make it in time, but undaunted, I heaved her over the wrought iron railing to get her onto the grounds; once there, she was able to get through the front door and into the building. Eilish departed soon afterwards for her holidays in Cork.

Eilish spent Christmas with her family in Cork, Ireland. She would take the night ferry to cross the channel, which was very uncomfortable due to the rough seas. I've never used the ferries myself, but I seem to remember her mentioning Fishguard to Rosslare. This might mean a train ride to Fishguard and ferry crossover to Rosslare, then a bus or coach to Cork.

BERTRAM MILLS CIRCUS
at Olympia
Wednesday 21st December 1955

Royal Performance
in the presence of
Her Majesty the Queen
and
His Royal Highness The Duke of Edinburgh

IN AID OF
THE LONDON FEDERATION OF BOYS' CLUBS
Sponsored by the Variety Club of Great Britain

DOORS OPEN AT 6.45 P.M.
CIRCUS STARTS AT 8 P.M.

BLOCK	SEAT	
4	64	£1. 1s.

TO BE RETAINED

PATRONS MUST BE SEATED BY 7.40 P.M.

First Date with Eilish Hannigan 1955

I spent Christmas visiting my sister Margaret in Oldbury, near Birmingham. Her husband had died not long before, and Margaret was left supporting her son, David, who was about five years old. I gave him an electric train set as a Christmas present, and enjoyed assembling it and playing with him. He seemed happy.

During my visit, I happily showed Margaret the family photo albums that Godmother had given to me, and instead of a pleasant sharing of good memories together, I was shocked at her reaction. "I want these albums because it is mostly about Papa, Mummy and me in North China," she demanded!

I was aghast at this reaction, and admittedly, I would have shared some of the photos with her, but I felt her demand was unfair. There were other photographs of Julie and me when we were babies and very young, but that was ignored. Anyway, I removed what I wanted, and left the photograph albums behind when I returned to London.

I then contacted Marjory Kirby and was invited to visit her and her family in Saffron Walden in Essex. Marjory was the social worker who had befriended Godmother when we first arrived in England after the war in 1945. They had kept in touch over all these years.

I came bearing a gift for her daughter Ann, which I had purchased from Harrod's in London; Ann was about five years old, but Marjory was very impressed because the gift wrap showed where it had been bought. I gave Marjory all the news I had about Godmother and Harry, and we spent hours reminiscing about how we had first met. After one night as her guest, I returned to London and Mrs. McGee's.

When Eilish returned from Cork we started to go out together. I would meet her after work at the Moorgate Underground Station close to her workplace at M&G Reinsurance Company. We would have dinner at nearby restaurants, and it was then, I started to learn a little bit about her family members and that we were about the same age. Her father was retired from work, and two of her sisters were in nursing. I was totally embarrassed when I tried to explain something about myself; it was so complicated that, as usual, I didn't know where to start. So, instead, I concentrated on the immediate past and my aspirations for the future and decided that I didn't know her well enough to go into any further details at that time.

In the New Year, Eilish and I didn't see much of each other during the week, mainly because I was occupied with my studies at night and attending classes during the day. She was back at work, so it was only on weekends that we could get together.

It was at this time that Eilish introduced me to her cousin, Katie Belton. I imagined that Eilish was seeking a second opinion from one of her family members as part of our relationship development, and I think I was accepted. In any event, I invited them both for dinner one evening to a swanky restaurant in the posh Piccadilly Hotel. The ladies were suitably impressed and especially marvelled at the dessert of flambé baked Alaska, which they never had before.

Eilish knew London well and was a delightful guide in showing me the highlights of the city. I don't know how many bridges we crossed, or how many parks we strolled through, all most fascinating. Museums and galleries were particularly interesting to me. It wasn't long before Eilish wanted me to meet her uncle, Mick Sheehy, living in London.

Mick lived in Hayes, not far away from his construction offices near the airport. I immediately got along well with Mick and his English wife, Christine (Chris). They had a charming family of four children: Barbara was a toddler, twins named Margaret and Brendan and the eldest was Eddie. They had a comfortable, modern home in a residential district, with nearby parks where the children could play.

Then for some reason, Eilish decided to leave the hostel in Spanish Place and move in with Mick and Chris in Hayes. The only reason I could think of was that this would reduce her living costs, but it would add a great deal of travelling to her work across London. Worse still, it would be a long trip for me to visit with her from Mrs. McGee's, because at this stage, our relationship was becoming more intense. I solved one of my problems by moving to Hounslow, which was not far from Hayes, but to get to college, I had the same inconvenience as Eilish; that was not going to be for long because I was near the end of my course anyway.

Mick and Chris accepted my frequent visits to their house to see Eilish on weekends. Before long, I started to feel as if I was part of the family. I often accompanied Eilish and Chris when they were taking the young children to the park and had fun playing with the children. I noticed that Eilish was well-accepted by her little cousins, and that pleased me.

Eilish often babysat her little cousins while their parents were shopping or out for the evening, and this gave us a wonderful opportunity to be alone with each other. Mick used to tease the two of us, as he could see in what direction our affection for each other was going. I think he approved of me.

First Visit to Cork

Our friendship was getting warmer now, especially with Eilish's willingness to share me with members of her immediate family. This was when I thought it would be nice to visit her family in Ireland, to which she readily agreed. It was decided she would go ahead, and I would join her later in Cork. It was during the Easter holidays, and the An Tostal Pageant was in full swing. Eilish found lodgings for me at a B&B run by a Mrs. Ryan, a kind, elderly lady.

I was warmly welcomed by her father, Thomas Hannigan, and his wife Catherine in their home called St. Ann, located in Marble Hall Park. This was where Eilish grew up with her three younger siblings. Eilish had a good relationship with her parents, and it showed, which was in such stark contrast to my own experience.

As I had with Eilish, I again had difficulty explaining who I was. Over tea, they jokingly wondered whether they would see a 'Chinese man from Hong Kong' appearing on their doorstep. We all had a great laugh over that!

In the little time I had allowed myself, Eilish took me around her hometown to see some of the local sights. This included the main street (I think she called it The Parade), and along the river walk. We walked passed Ryan Soap Factory, where her dad used to be a director before his retirement. Of course, the Irish with their interesting lilt to the language added to my fascination.

Because of my class schedule, I left Cork before Eilish. I took a train to Dublin to make connection with a flight to London. Having to wait for my flight at the airport, I telephoned Paddy Thomson, a former shipmate of mine, at his home, as I knew he was also on long leave from Jardines. His sister answered the phone and gave me the number of the pub where he could be reached. When the publican answered my call, I asked for Paddy, as that was the only name I knew. With a roar of laughter, the publican said to me, "Begorah man, which Paddy would ya like me to produce for ya?" Realizing how stupid I was, I hastily described what Paddy looked like, and learned for the first time that his name was George—not very Irish, if you ask me. Talking to Paddy afterwards gave us both a good laugh; unfortunately, it was not possible to meet with each other.

Final Examination Success

On April 23, 1956, I was notified that I had passed the Ministry of Transport (MOT) final examinations and was awarded a Certificate of Competency as First Class Engineer of a Steamship. At 23 and a half years old, I would be the youngest chief engineer in Jardines. Upon receipt of this information, Captain Bernard Smith, Director of Jardine, Matheson & Co. in London, was authorized to extend my long leave an extra month. This extra time delighted both me and Eilish.

I was now ready to seriously consider marriage with Eilish. There were some enormous logistical challenges to consider which we would toss about deeply in the short time remaining between us. That evening, on our way in a taxicab to Soho for dinner, I proposed marriage to Eilish. I thought if there were any uncertainties, we could resolve them over dinner. Her response was a positive hug as she whispered my

Dinner at Veeraswamy Restaurant in Lower Regent Street in London with Chris, Mick and Eilish. To celebrate my successful First Class Marine Engineering examinations. 1956

At a dance in Cork. May 1956

Getting acquainted in the parks of London, 1956

name. At last, I felt that perhaps now I would have a purpose in life, not that I had any idea what it would be like.

Dreamily, we both entered a nice Portuguese restaurant in Soho and, because it was in the middle of the week, the restaurant was practically empty, except for two older gentlemen in business suits sitting quite some distance from us. Eilish and I were so elated that we were oblivious to our surroundings; I can't even remember what we ate. We were very surprised when a waiter came to our table with a bottle of wine, courtesy of the two smiling gentlemen nearby. I guess they could see that we were terribly in love, and after all, it was springtime!

The next thing on our minds was to announce to Mick and Chris that we considered ourselves betrothed and invite them to celebrate this event with us, as well as my success in the examinations. That weekend, I hosted dinner for the four of us at Veeraswamy restaurant in Lower Regent Street, where the specialty was curry. The ladies were beautifully dressed for this special occasion, and photographs reveal a happy group. Chris and Mick took to the dance floor, and we also followed, dreamily holding each other.

Second Trip to Cork with Eilish

That spring Eilish and I travelled together to Cork for me to request Eilish's hand in marriage from her father. It was an exciting time for both of us, but for me it, was also scary, wondering how I would manage all these extra responsibilities in due course; but joy soon overcame all these thoughts.

Marie, Eilish's sister, and her fiancé Ken, came to the airport to meet us and drove us to Marble Hall Park. It was a great reunion with Tom Hannigan and Eilish's mother. After tea that evening, I approached Tom in the front parlour and made my announcement. I think he was impressed by my accomplishments and consented to my request to marry his daughter. Eilish was with her mother in the nearby kitchen, waiting for the result. Soon, I came out beaming from ear to ear, and hugged Eilish and her mother, thankful that formality was successfully behind me. From then on, Eilish and I could seriously plan, even though it was mostly in our dreams at this stage.

That weekend, Ken very kindly drove the four of us in his Ford Prefect to different parts of the nearby countryside. I wish I could remember where he took us, but there were lakes and a derelict castle where the wind was blowing fiercely. I think Eilish and I went on our own to Blarney Castle later, where she insisted that I kiss the Blarney Stone. It was a wonderful week we had together, constantly talking about our dreams, and about the kind of life awaiting Eilish in Hong Kong.

We stayed a couple of days in Dublin on our way to London because Eilish wanted to introduce me to some of her favourite aunts. Aunt Cissy and Aunt Joan, both religious nuns in a convent, gave us a great welcome. They were so sweet to us, and they couldn't help remarking how young we both looked. In Raheny, Eilish stayed overnight in the home of Aunt Agnes; I checked into the Gresham Hotel. I met Aunt Agnes' husband Donal McCarthy, and their little daughter Maire. I was warned beforehand that Aunt Agnes had a crusty reputation, but not to be alarmed. I was pleasantly surprised at the warm reception I received at our introduction. Eilish told me later that Aunt Agnes was quite impressed with me!

Back in London, we found the spectre of inevitable separation weighing heavily on our spirits. We were conscious how it was affecting us both, and sought comfort in our closeness with each other as much as possible. I think that the impending separation was going to be harder for Eilish; as for me, I was going back to a job that I had to adjust to and would keep me occupied. So, to lighten our spirits, I decided that a day trip to Paris would be an excellent remedy, to which Eilish heartily agreed. This would be a first-time experience for both of us, and it would add significantly to our memories.

Day-Trip to Paris

It was the Whitsun holiday weekend, and the atmosphere was quite festive. In addition, spring was blooming all around us. On May 19, we flew out of Croydon, also known as the London Airport, and arrived in Paris at Le Bourget Airport. Because of our time constraints, we made full use of the available taxi service. Highlights of our visit included a stroll through the Luxembourg Gardens; Notre Dame Cathedral, which also reminded us of our own Westminster Abbey in London; and the nearby Sainte-Chapelle, a former royal chapel with beautiful stained glass windows.

Crossing the Seine River, we walked the Left Bank for a while, then popped over to the Musée d'Orsay. We took a jaunt to Montmartre, where we had some lunch, and for dessert, an ice cream cone as we walked along the cobbled streets.

Weekend in Paris - view of Paris from Montmartre

The view of Paris from Sacré Coeur was stunning, and before leaving Montmartre, we checked out the Toulouse-Lautrec museum. As it was getting late in the evening, we ended the day by having dinner at the 'restaurant in the sky'—the Eiffel Tower. The night view of Paris was brilliant and so romantic. Near the end of the meal, Eilish suddenly exclaimed that we had to rush to the airport right away, otherwise we might miss our flight. In great haste, I paid the bill, but because we had no time to have our dessert, I received two vouchers from the restaurant to come back and finish our dessert later. I still have those vouchers!

During the flight back to London, we were both rather quiet, thinking about the separation to take place next week. We squeezed hands in support of each other and returned smiles bravely to those around us.

No date had been set for the wedding, but it was understood that I would continue with my employment with Jardines in Hong Kong, which would mean a temporary parting. The separation was going to be hard for both of us!

Last Days in London

I well remember how we spent the last few days together in London. Spring was in the air as we walked through the parks holding

RESTAURANT
"En plein Ciel"
DE LA
TOUR EIFFEL

par ce chèque ANDRÉ PIGNARRE
se fait un plaisir de vous
rembourser l'ascenseur!

Tél: INVALIDES
19-59 et 44-67

POUR VOS DÉPLACEMENTS
UTILISEZ
LES CHÈQUES DE VOYAGE
B. N. C. I.

Chèque N° LA445.216 B.P.F. -100-
BANQUE NATIONALE POUR LE COMMERCE ET L'INDUSTRIE

PAYEZ CONTRE CE CHÈQUE Cent Francs

A l'ordre de au porteur

le 19 MAI 1956

CE CHÈQUE EST PAYABLE DANS
TOUS LES SIÈGES DE LA
BANQUE NATIONALE
POUR LE COMMERCE ET L'INDUSTRIE

André PIGNARRE (Restaurant de la Tour Eiffel - PARIS)
Auxerre c. c. N° 2220

Weekend in Paris –
top: enjoying an ice ream cone
centre: Luxembourg Garden.

Two vouchers from the restaurant at the Eiffel Tower dated 19 May 1956 because of our hurried departure to catch a flight back to London.

hands, sitting on the park benches, listening to the sound of ducks in the ponds. It was exquisite happiness, but tinged with the shadow of our forthcoming separation, and to be separated by an unimaginable vast distance.

On a serious note, I tried to explain to Eilish the lifestyle she might expect in Hong Kong, and the prospects open to me from Jardines. Early retirement in 1986 at 55 years of age with full pension and a provident fund designed for purchase of a home in England at the time of retirement. From what I had already experienced and seen, senior employees and their families were provided attractive accommodation in preferred districts; many eventually resided on the Peak. Social life was renowned in expatriate circles, to which we would belong, and eight months paid leave every four years to Ireland or England on full salary, less the Far East cost of living allowance. Altogether, not too shabby!

The parting was not easy for either one of us. We buoyed each other up by talking about the future and made all kinds of promises to keep in touch. At the time, we weren't sure when the wedding would take place, or where—in her hometown or in Hong Kong. We hugged and kissed each other goodbye That was on May 24, 1956.

Events were now beginning to crowd into my life at a pace which left me gasping. I relished giving my all to my significant partner in life, but I wasn't sure if I knew how to manage it. I dare not let her know how inexperienced I was! I desperately needed to talk about my concerns with somebody I could trust, but with whom? I had no close confidants at this time of my life!

My emotional state was quite high as I gathered my belongings in Hounslow and headed by taxi to the nearby airport, pondering the enormous commitment I had made to Eilish. Thank goodness the business of departure had kept me fully occupied.

The flight was by BOAC in a turboprop Argonaut aircraft, and the first stop was in Rome (this was the actual aircraft that transported Princess Elizabeth and her husband on their honeymoon to Kenya). In the terminal building, I saw Anthony Steele and Anita Ekberg, who were in transit to a honeymoon destination. Then followed Bahrein International Airport and Dum Dum in Calcutta, finally return to Hong Kong landing at Kai Tak on May 25.

CHAPTER 13: GLORY YEARS

Promoted to Chief Engineer with Jardines

1956 – 1959

The great difference between voyages rests not with the ships, but with the people you meet on them.

- Amelia Barr

John Green, the superintendent engineer, warmly welcomed me when I reported to his office in Hong Kong on May 25, 1956, congratulating me for attaining the First Class Certificate of Competency.

I was placed on reserve, and accommodation was arranged for me at the European YMCA in Tsim Sha Tsui, pending my assignment to a ship. The next 10 days were a welcome respite after the long flight from London, and it also provided me the opportunity to visit with my sister and her family, and we talked about Eilish and our future plans.

On this tour, I would serve on the *Eastern Star, Eastern Saga, Hew Sang* (for the second time), Tak Sang, and Hew Sang again (third time, and last ship).

Eastern Star: June 1956 – January 1957

This ship was on a trading route for Australia > Hong Kong > Japan—a beautiful-looking ship with smart lines.

I joined the SS *Eastern Star* on June 5, 1956, as the second engineer, two days before departure from Hong Kong. The ship was in the Hongkong & Whampoa Dockyards (HWD), completing the first Special Survey since being built. The engineers listed were new to me, except for Allan Culpepper; Bill Rowe was the chief engineer; Keith Highfield; Bob Allan; Bluey Curnow; Jim Fontaine; George Adam; Eric McCall; and Steve Chan.

At the time, HWD was under considerable pressure to have the ship ready to sail on schedule. Bill Rowe and I worked long hours together during the last two days (and nights) with the dockyard personnel verifying and approving the work being done. Adding to the pressure, passengers were boarding the ship in the dockyard at the very last minute. I can still recall that, by the time we sailed out of Hong Kong, I had put in nearly 30 hours of continuous work without any sleep! I felt like a zombie at the end.

This was an impressive ship, similar in class to the *Eastern Queen*, but slightly smaller at 6,523 tons. The ship was classified as a cargo liner, with 12 luxurious staterooms for passengers, and equal in quality for officers, all with air conditioning. The ship was built by Harland and Wolff of Belfast in Northern Ireland in 1951, and propulsion machinery consisted of Parsons triple expansion steam turbines rated at 7250hp. High pressure steam was supplied from two water-tube boilers, and electricity from four diesel generators. I was really looking forward to this new experience.

I was very thankful that on this voyage, we had surplus watch-keeping engineers, and with the chief engineer's consent, I did not have to stand a watch and was thus able to catch up on lost sleep. Being free of watch-keeping duties, I could then also leisurely acquaint myself with the ship's machinery and have continuous consultations with the chief engineer—a most agreeable situation.

Eastern Star in harbour c1956
Built in Belfast by Harland & Wolf 1951 Air-Conditioning was also available in all accommodation quarters, as well as in the public areas.

Southbound – Across the Equator

The incident that shouldn't have happened because of all the work done at the Special Survey in HWD, did happen! This could probably be blamed on human error at the time. While at sea in the middle of the ocean, power from the two diesel generators failed, one at a time. The standby generator also failed shortly after start-up, and the fourth generator was out of commission, as it was being serviced at the time.

With the ship drifting in the ocean after it had come to a complete stop, Bill Rowe and I joined the operating engineers in the engine room to determine the cause of the outage. Fortunately, it didn't take long to be identified, and it was easily fixed, much to the captain's relief.

Suspecting the problem was associated with the diesel fuel delivery system, we quickly located the master fuel filter hidden underneath the steel floor plates in the engine room; it had never been serviced since the original installation. It was quickly dismantled and properly serviced, and before long, full power was restored and the ship was once again underway.

'Baptism on the Line', also called equatorial baptism, is an initiation ritual sometimes performed as a ship crosses the equator, involving water baptism of passengers or crew who have never crossed the equator before. The ceremony is sometimes explained as being an initiation into the court of King Neptune.

For this occasion, the ship's crew had erected a makeshift pool on the deck, and one of the deck officers was dressed up to represent King Neptune. Victims selected were only the first-timers who were dunked into the pool, and when thoroughly soaked, would be plastered with flour to make them look like ghosts. King Neptune, with a crown, would hold court, and at the end of it all, each victim would receive a certificate signifying their successful induction into the court of King Neptune. Thankfully, I was spared this indignity because at the time, I was in the engineering office with the chief engineer. The passengers took it all in good humour.

When the ship arrived in Brisbane, also in port was the *Eastern Argosy*, on her maiden voyage from the UK, now northbound for Hong Kong. I heard afterwards that the chief officer was my old shipmate, John Stormont, from the *Chun Sang* days when he was the second officer. I sure would have liked to tell him all about my fiancée, but I didn't know he was there. He probably didn't know that I was on the *Eastern Star* either. Ron Learoyd joined the *Eastern Star* from the

Eastern Argosy. I had also sailed with Ron on the *Eastern Queen*, so I knew him well.

Meeting Godmother and Uncle Harry in Sydney

It was in Sydney that Olga and Harry Blake came on board to have dinner as my guests on the ship. Devoid of passengers, we were almost the only ones in the dining room that evening, and as required by regulation, I was wearing my uniform. I was also the most senior engineering officer on board the ship at the time, as Bill Rowe, the chief engineer, had taken time off the ship to be at home with his family.

I took this opportunity to announce my betrothal to Eilish, and of our intentions to be married and settle down in Hong Kong. Because of this, I also advised her that the monthly remittance of HK$200 to her bank account would now be terminated due to my new responsibilities.

After dinner, we retired to the lounge, where we enjoyed coffee and after-dinner liqueurs, and Harry had his smoke. We continued talking about Eilish and how I was missing her. Godmother was very interested and impressed, especially when I showed them photographs. It was thus that we passed a very pleasant evening.

I was aware this was an opportunity to question Godmother about many unanswered questions I was carrying from the past, to which she might have answers. How well did she know my parents? Did Mother leave any assets behind when she died, and who had custody? Why didn't she rescue me from DGS when the war started? Why didn't she embrace me in Stanley Camp when she must have recognized me? Why didn't she acknowledge the letters I sent to her in Stanley Camp? How did she acquire some of the family assets so many years after the war? Maybe I was still unsure about my relationship with my godmother; perhaps I didn't know how to broach the subjects to her, or I was reluctant to cause her embarrassment. I couldn't fathom what was holding me back at the time!

When the evening ended and they were preparing to depart, I convinced myself that when I next visited Sydney, I would have enough courage to deal with these and any other outstanding questions. This would haunt me for the rest of my life. Unfortunately, I was never to see them again, as shortly afterwards, they moved away from Sydney, although Godmother would exchange correspondence with Eilish from time to time before she died in February 1977; Harry died in March of 1966.

§

With the married Australian officers ashore with their families, we were short-staffed on board the ship. This kept me busy during the day, managing maintenance work in the engine room with my engineers. We were only four engineers and our regular Chinese engine room workers. I was so occupied that I never once set foot ashore in Sydney on this visit, not even to look around! However, the bachelor officers, who were 'in the know', were not without amusement after hours and on the weekend.

The bachelor officers who come to mind were Ron Learoyd (deck officer), George Adam (engineer), Bobby Allan (engineer), Eric McCall (engineer) and a few more. Many of them had contacts ashore from previous visits, and it wasn't long before they were entertaining female company on board the ship. The source of this female company was usually nurses from the local hospital, and it wasn't unusual for their matron to chaperone her young charges, at least on the first visit.

One evening, I had decided to retire to bed early, after a hard day's work in the engine room. I was quietly re-reading some of Eilish's letters, and in the background down the hallway, I could hear a developing party in somebody's cabin. Well, later that night this party "poured" into my room, just when I was falling asleep. So, with a beer handed to me, I sat up in my bunk, accompanied by two very attractive young ladies, and joined the party. When one of these girls found out that I was betrothed, she wailed out loud and said, "How come all the good guys are taken?" I think her name was Ann, and she was sharing a flat in King's Cross with her friend.

Melbourne – Northbound to Japan

I don't remember much about Melbourne, except that it was host to the 1956 Summer Olympic Games at the time. One evening, Keith Highfield and I went ashore to the Opera House to attend a performance. The ship took on cargo and perhaps some passengers and turned around to head north. We were back in Sydney for a few more days before proceeding to Brisbane in Queensland.

From Brisbane, the ship made a stop in MacKay, further north up the coast, to load live beef cattle for Manila. For this purpose, temporary wooden pens were constructed on deck, and the herd was accompanied by a certified veterinarian. This included high-pressure daily washing of the decks around the cattle pens. In due course, the unloading of live cattle in Manila proceeded without incident. The next stop was our home port in Hong Kong.

St. Teresa's Church and Kowloon Riot

In our last exchange of letters, Eilish had graciously agreed to have our wedding celebration take place at a church in Hong Kong. This was most practical for me and most convenient for my employer. The understanding was that she would stay with my sister Julie in her apartment at Nairn House until the day of the wedding. Eilish had previously met Julie and her family in London, when they were on long leave at the time; they had brought their amah with them on that trip to care for their two pre-school children.

My preparations included the selection of a church to conduct the marriage ceremony. I was quite scared by the prospect of such a life-changing decision, and again, I had no idea what impact this would make on me. It would have been of enormous benefit to be guided by family support, but this was not available to me. My biggest and only support would eventually come from my betrothed, to whom I would owe so much!

So, on October 10, 1956, when the ship was in port, I made my way by bus to St. Teresa's Church on Prince Edward Road. I explained to the parish priest my plans for our marriage ceremony for next year. I was made aware of all the church requirements, which I conveyed to Eilish. The date was tentatively booked, as selected by Eilish, to be confirmed when the day was closer.

At the end of the meeting, I felt so relieved when I boarded a bus outside the church to return to my ship in the harbour, that I was unaware of a disturbance happening on the street the bus was driving through. We were in the middle of a full-scale riot!

By the time we reached Mong Kok, I was surprised to note that I was the only passenger on the bus. Due to the mob milling around on the street, the bus was crawling along quite slowly. The Chinese bus driver suggested that I refrain from exposing myself through the bus window and urged me to hide myself underneath a seat. It was a wise decision, as before long, a few of the agitated rioters tried to board the bus, apparently searching for foreigners to assault. They couldn't make entry onto the bus, and peering through the window, they were unable to see me crouched underneath a seat, and lost interest in the 'empty' bus at that point. Did that save my life? Who knows!

As the bus picked up speed going south on Nathan Road, I looked behind me through the rear window and could see that the mob had trapped two uniformed British soldiers from the nearby barracks outside the entrance way to the Mong Kok Cinema and were pelting them with rocks. By the time the bus reached the Jordan Road intersection, I could see the charred remains of an automobile in the middle of the crossroads, within which the wife of the Swiss consul had been killed.

I was quite relieved to get off the bus at the Star Ferry Terminal in Tsim Sha Tsui, and boarding a walla-walla (water taxi), I made my way back on board the Eastern Star in the middle of the harbour. It wasn't until much later that I heard about the ferociousness of the riot during which the governor of Hong Kong had called upon the military to assist the police to quell the rioters. Sometime later, I read about the riot from the following newspaper report:

> *"On 10 October 1956, riots broke out in Kowloon, lasting for three days, with many casualties. The riots, built on the continued tension between the Communists and the Nationalists, brought terrible scenes of violence that resulted in 60 deaths and 443 people injured. It remains the riot with most casualties in Hong Kong history."*

I do not recall any discussions made by the officers and crew about the riot in Kowloon, and perhaps the gravity of the incident only surfaced several days after we had departed on our way to Japan. Typical ports of call ahead of us were Nagoya, Yokohama, Osaka, Kobe and perhaps Yokkaichi, which was the import port for Australian raw wool.

Excursion from Kobe to Takarasuka

Together with Keith Highfield and Bill Rowe, I had an interesting experience on this visit to Japan. It was in Kobe that we decided to take a day off to explore the nearby town of Takarasuka, 'City of Music', famous for its opera house and all-girl revues.

It was a fine, hot day when we set off in a taxi through the neighbouring countryside, when the taxi was stopped by a collapsed bridge over a stream. We decided to pay the taxi driver and continue walking towards our destination along the quiet country road in the heat of the afternoon. The odd farmer and women waved to us as we trudged past small farm holdings and came across a pool fed by a small creek; we thought it would be a nice idea to cool off with a dip.

I never knew skinny dipping could feel so luxurious, as we soaked in the pond. The three of us had a great laugh, and before long, fully refreshed, we continued walking on our way to Takarasuka. As we got closer, we started hearing sounds of music wafting over the air towards us, and later discovered that it was coming from speakers attached to the town's streetlight posts.

The show at the Opera House was a musical titled Springtime in the Rockies, with all-girl performers dressed in the famous red serge of the Royal Canadian Mounted Police (RCMP), some mounted on live horses. At the end of a very satisfying day, we went to the railway station and took a train back to Kobe to rejoin our ship.

§

I remained on the Eastern Star for one more trip to Australia. As can be imagined, my mind was becoming more and more preoccupied with my future change in status. Correspondence between me and Eilish was agonizingly slow, not only because of 'snail mail', but because of the mobility of the ships I called home.

I served on the ship for another two months, making a round trip from Hong Kong to Sydney and Brisbane, and getting off in Hong Kong without continuing to Japan. I signed off the Eastern Star and was placed on reserve in Hong Kong for 21 days.

On Reserve in Hong Kong: January, 1957

During this period, I made a formal request to the superintendent engineer, John Green, for permission to change my single status to marriage status. By now, Eilish had confirmed November 6, 1957, as her date of choice for our marriage in Hong Kong; this was readily granted. I was given complete instructions to convey to Eilish about company travel arrangements.

Eilish and I had a happy telephone conversation, and she seemed pleased that plans for our reunion were progressing so well. She told me that the tailoring of her wedding dress was in progress back home in Ireland. She would quit work soon and prepare her trousseau to bring with her to Hong Kong.

My sister Julie and her husband were also delighted for me and assured me that they would look after Eilish when she arrived in Hong Kong. Final details were still to come, but for now, I was in seventh heaven!

MV Eastern Saga: January 1957 – April 1957

MV *Eastern Saga* entering Sydney Harbour in Australia c1957.
The principal service of the ship was the transportation of cargo, but it also had accommodation for 12 passengers.

At the time, the *Eastern Saga* was the only vessel in Jardines equipped with diesel-driven propulsion machinery. It was John Green's intention that I should gain experience with this class of machinery, to prepare myself to write the examination for a diesel endorsement to my Chief Engineer's Steam Certificate. The MOT prerequisite was six months of operating experience. Therefore, I was keenly interested in the machinery on board this ship.

Classed as a general cargo liner, there were 12 first class passenger staterooms. Accommodation was also first-class for ship's officers. The trade route for this voyage was New Zealand > Australia > Hong Kong

The *Eastern Saga* was originally named the Esmeralda. She was built in 1944 in German-occupied Holland and had a gross tonnage of 6,631. She was acquired by the MOWT at the end of the Second

World War, in May 1945, and went into service under the British flag under the name of Empire Wye. Jardines purchased the ship in April 1947. Keith Highfield and 'Two-Gun' Thompson were there to take delivery and supervise the overhaul in a Hamburg shipyard, and the now *Eastern Saga* sailed to the Far East to join the rest of the fleet.

I was awestruck by the massive amount of iron when I first set foot into the engine room; the height of the diesel main engine from the floor plates was over 25 feet. All pressure ratings and temperatures were measured in metric. Also, most of the signage in the engine room was in German.

The main propulsion engine was a MAN two-stroke diesel with eight cylinders rated for 4,000bhp, plus two scavenge pumps. There were four diesel electric-driven generators, with two connected to air compressors.

The captain was Duncan Kinnear; the chief engineer was Douglas 'Two-Gun' Thompson and the Second engineer was Colin Davidson, an Australian new to me. I had sailed under Duncan Kinnear before and had known about 'Two-Gun', an old timer with the company. So, I reckoned I was in good hands to learn from this new experience.

My watch at sea was 12:00 to 16:00, and midnight to 04:00. I was also responsible to test the main engine diesel fuel injectors after they had been overhauled by independent contractors. The on-board maintenance shop included facilities to do this precision testing. I also found it interesting to work with the second engineer on the main engine crankshaft and cross-head bearings, especially having to bodily enter the crankcase.

When entering or leaving port, there were always two engineers at the manoeuvring station in the engine room to respond to telegraph signals from the bridge. It was so neat to feel the power of the engine through the throttle control at times of manoeuvring, especially when the engine reversed rotation.

The round trip started and ended in Hong Kong, via Singapore, Auckland, Sydney and Brisbane.

I had no interest in getting off the ship and doing any exploring in Auckland and vicinity, our first port of call; or on other stops along the Australian coast. My main preoccupation was mentally preparing for marriage, and for the anticipated examinations to earn my Diesel Endorsement sometime in October at the end of six months.

'Two-Gun' Thompson was interested in my studies on board the ship and was always ready to give me advice. He used to peer through the porthole to my cabin from the outside, and mischievously say, "Big brother is watching you!" It was well-known that he spent most of his entire long leave every four years on safari in Africa. He had quite an arsenal of weapons in his cabin, and trophies from previous safaris. To keep in practice, he had a rig set up at the rear of the ship to fire off clay pigeons to practise shooting.

I didn't know it at the time, but this was to be my first and last voyage on the *Eastern Saga* as a serving officer. In Hong Kong on April 28, 1957, I was removed from the ship and placed on reserve. Regretfully, I never did complete my service time to permit me to sit for the Diesel Endorsement Certificate. However, the positive aspect of this move was that I was promoted to my first posting as chief engineer on my next assignment.

SS Hew Sang: May 1957 – October 1957

Hew Sang fully loaded with timber for Japan. c1957

This was my second appointment to this ship, and my first as chief engineer. How did I feel about this? I was happy to be back in familiar surroundings, but my main pre-occupation at the time was my forthcoming marriage in November. Not only that, despite keeping in constant communication, I couldn't help but wonder what changes I might see in my Irish Rose when she stepped foot in Hong Kong, and we came face to face again after 18 months of separation.

As noted on my earlier assignment to Hew Sang, the trade route was Hong Kong > British North Borneo > Japan > back to Hong

Kong. Our principal cargo was wooden logs from the forest of Borneo, delivered to Japan. Returning to Hong Kong, we never had a full load of cargo. What we did have were bags of cement and maybe some building materials.

The crew included Captain Grieve, normally referred to as the master of the ship; I remember he enjoyed his wine, Chablis being his favourite, which he used to buy by the case. George Hurford, my second engineer, originally came from Durban in South Africa. George came to us from the Union Castle Line; I found him knowledgeable and likeable. Bill Povey and Jan Jensen joined the ship at the same time. Povey was the mate, or first officer, and Jan was the second mate, or second officer. As for Dave Wilson, our third mate, also known as 'The Roadrunner', there is quite a story about him later.

At age 26, I had no idea what was expected of me as chief engineer, apart from having observed at close quarters how all my former chief engineers had managed their responsibilities. The master of a ship always maintains a certain distance between himself and his officers based on two criteria: maturity due to age and responsibilities due to rank. Chief engineers have the same criteria, except that they report to the master of the ship, because the master represents the owners of the ship. The master is somewhat aloof from his officers and the rest of the ship's crew, except for festive occasions. The engineers and deck officers got along well with each other, principally because of compatibilities due to age. With masters typically aged in their 40s, and me 26, one could expect that my social tendencies were more in harmony with officers similar in age to me.

Dave Wilson and the Python

We were at sea on one of our trips to Borneo, and at a social gathering, the officers got into a discussion of shipboard pets. Cats weren't uncommon, and we had one on the *Hew Sang*. Frank Lee on the *Wo Sang* had two canaries in a cage. But Dave Wilson, our third officer, held the theory that a snake was the ideal pet to have on board ships. The consensus from the officers was no matter how useful or entertaining a snake might be, it is highly impractical to have a snake loose on board a ship. In the ensuing argument, nobody denied that snakes do have certain useful qualities, but Dave refused to acknowledge the unpredictability of a snake's behaviour and the impracticality of stowing a snake on board a ship as a pet.

Shortly after we arrived at Sandakan, during a business meeting with the company agent, John Povey, the chief officer, was informed that the local villagers had recently found a python snake in the nearby jungle. With this knowledge, an idea was born! Without Dave Wilson's knowledge, the ship's officers arranged with the villagers to deliver the python on loan to the ship for an experiment. Apparently, the snake had recently gorged itself on a meal and was considered docile for the time being.

Most of the officers, including Dave, were gathered in the ship's saloon at mid-ships, when the eight-foot python was delivered in a wooden Carnation box. Entering Dave's cabin, the wooden box was emptied on his bunk, and there lay the docile, coiled python, and the door carefully closed. Beforehand, our cat, the ship's mascot, had been stowed safely away in one of the crewmen's quarters.

Meanwhile, the rest of the officers were enjoying themselves in the saloon when John Povey, the chief officer, requested something from Dave, which he knew was inside his cabin. Of course, all obliging, Dave bounded down to his cabin at top speed (hence his nickname, 'The Roadrunner'). In the meantime, some of the officers were outside, peering through the porthole of the cabin, waiting to see Dave's reaction. Dave flung open the door, rushed over to his desk with hardly a pause and retrieved the object requested, and without batting an eye, returned to the saloon upstairs. He didn't even notice the snake on his bunk!

The next move was to uncoil the snake, and using its length, stretch it across the path that Dave must cross next time. So once again, John Povey made a request, and once again, Dave bounded down the stairs and into his cabin. With great delight, the spectators outside peering into his cabin saw Dave skid to a stop as his eyes fell on the snake stretched across the floor and in his way. His face had turned as white as his uniform as he beat a hasty retreat!

Back in the saloon, the officers poured a stiff drink into Dave to calm his nerves. He didn't say very much, and perhaps he realized that the officers were just having some fun at his expense. In the meantime, one of the villagers had gathered up the snake to remove it from the ship. While we were laughing and talking about the python, somebody at the entrance threw into the room a wooden Carnation box, and everybody immediately jumped up onto their chairs, expecting to see a python fall out of the box, but it was empty—the trick backfired against the perpetrators.

Dave never again talked about the suitability of snakes as pets on board a ship. I don't know how long that snake grew in life, but it got longer every time the story was told. By the time the tale had made its way around the rest of the ships in the company, the snake had grown to 15 feet. It took Dave a long time to live down the python story.

Rescue at Sea

Before leaving North Borneo with a full load of logs for Japan, we often stopped at Tawau to take on fuel oil for the boilers. On this occasion, we were the second ship in line at the Shell Installation. Our turn to load came at about 1800 hours. Dusk was falling by then, and by the time we were finished bunkering, it was almost fully dark when our ship sailed off.

It was a very pleasant evening, and several of us officers were enjoying dinner on the open deck when we heard a faint voice out in the dark ocean; it sounded like a call for help! We heard it again, and the captain slowed the ship so we could investigate.

We slowly got closer to the voice—unmistakably a male! We tossed a buoy into the water and he said, "I see the light." There was some rhythmic splashing, as of someone swimming, and soon a lone figure clung to the buoy. We dropped out a Jacob's ladder, and a young man in bathing trunks clambered up to meet us.

The young man was an officer cadet from the Australian ship that had bunkered ahead of us and had departed first. That ship had a temporary swimming pool built alongside one edge of a well-deck. The cadet was alone in the pool that evening, as his fellow officers had all gone for supper and left him behind. Last out, this cadet had heaved himself out on the wrong side of the swim tank and fallen straight into the ocean. Most of the officers were in the dining room and the noise from the ship drowned out his cries. The young cadet had the presence of mind to position himself between the fading lights of his ship and the approaching lights from ours. He waited for about two hours, treading water the whole time.

After we had rescued the young cadet, our captain radioed ahead to advise the other ship that we had rescued one of their officers, and we arranged a rendezvous in mid-ocean to transfer him over. His captain and crewmates had noticed his absence at dinner but thought he was having a snack in his cabin in order to catch some sleep before going on watch at midnight.

They were astounded to learn what had actually happened and considered the cadet very fortunate to be rescued after two hours in those shark-infested waters. The story of the rescue appeared in a local newspaper in western Australia, where the young man was from, and his parents sent us expressions of gratitude for our part in saving their son.

Fun with My Speedboat

Becoming the owner of a 16-foot speedboat was a spur-of-the-moment decision, which perhaps I shouldn't have entertained. We were in Sandakan on one of the earlier round trips, and our local shipping agent told us over lunch that he had wrecked his speedboat on some coral reef and made a mess of the 60hp Mercury outboard engine. The hull wasn't too badly damaged, he explained, and being made of wood, it could easily be fixed. This information intrigued me.

The next day, I had a look at the boat, and when the agent said he would accept $100 for the boat, as-is, I bought it. With the help of the ship's carpenter, I easily patched the gash in the hull of the boat, and when next in Hong Kong, I purchased a new 18hp Evinrude outboard engine; I felt the original 60hp outboard was too powerful for this size of a boat. Captain Grieve had no objections to me stowing my boat on the Hew Sang.

The little speedboat was very useful for gaining access to pristine beaches on small islands. All we had to do was to get over the encircling coral reefs and beach the boat into the sand. On one expedition, John Povey, Jan Jensen and Wilf Briggs joined me, and we headed out with supplies to have a barbecue on one of those island beaches. We had two chickens, some sausages and quite a few beers.

We gathered material and pieces of driftwood to build a barbecue, which we set afire. Two of us fashioned a couple of skewers from nearby branches to cook the chickens, and the other one started to cook the sausages in the frying pan over the same fire. With this magnificent start, we decided to celebrate with some beer. All was going well while we watched the fire, until we were suddenly overcome by one of those frequent tropical flash squalls, which usually blows away within 15 minutes.

My speed boat salvaged from a wreck on the Borneo Coast near Sandakan. I am at the motor. c1957

Me (l) exploring an island with my boat near Sandakan, together with George Hurford and Jo Jensen.

For some reason, the grease in the frying pan caught fire and charred some of the sausages. By the time the squall had passed, we thought the chickens were cooked sufficiently, but we found out that they were not fully cooked because we hadn't removed the giblets wrapped in paper and stuffed inside the chickens. So, our first barbecue on a beach in Borneo was not a spectacular success, but we were saved by our supply of beer, and that gave us sufficient cheer. Before returning to the ship, we whiled away some time by waterskiing within the lagoon, which worked well if we lightened the weight of the boat by the removal of two bodies during skiing.

Most of the time, we used the speedboat for exploring nearby unknown islands that looked interesting. On one island, we came across a tribe of wild monkeys; as they didn't seem too friendly, we steered clear of them. On another island, we were surprised to stumble across a small fishing village located right on the edge of the beach. The inhabitants showed friendship with a lot of toothy smiles, but we didn't understand their language. However, using sign language, I managed to barter two paddles for my speedboat; pointing to two rough-hewed wooden paddles inside one of their canoes, which signified my intentions, and in response, they pointed to two packets of Lucky Strike cigarettes, and that sealed this foreign trade transaction to everybody's satisfaction.

Coming back to the ship anchored in the bay, I carelessly let my two paddles slip into the water. This is normally of no consequence, but one of the paddles was hewed from a tropical hardwood. It was denser than water, and immediately sank to the bottom of the bay. I eventually sold the boat to a fellow officer when I signed off the ship to get married.

Driver's Test in British North Borneo

I was anticipating that when Eilish and I became settled down in our married life, that it would be a good thing for me to have a driver's licence so that I could show her the sights of Kowloon and Hong Kong on my days in port. With this in mind, I had planned to take a driver's test in Sandakan.

Having never driven a motor vehicle before, I needed instructions and practice. Fortunately, the ship was in port for about seven or eight days. My instructor was the friendly cargo agent for the company; the vehicle at my disposal was a black Mayflower with a standard shift gear. I practised on about seven kilometres of paved road with one traffic light, attached to an overhead log chute straddling the main street. The light would turn red when the overhead log chute was in use to direct logs into the nearby bay.

Our local agent introduced me to the local examiner for a driver's licence. He was also the fire chief and chief of police, though I didn't know that at the time. He was very courteous and friendly, and after some preliminary questions, off I drove. I stopped at the red light above the log chute before proceeding; well done. Everything went smoothly, until at the end of the road test he directed me to back up into the fire hall. I hadn't practised reversing at all, but I completed reversing into the fire hall and stopped beside the fire truck without incident.

But horrors, I had stopped so close to the big red fire truck that my examiner could not open his door to get out of the car. He issued my driver's license anyway, and I invited the examiner on board the ship as my guest for dinner that night to celebrate my success:

Driver's License. Colony of North Borneo Licence No. S1917 – Issued 27 July 1957.

A month later, I took a driver's test in Hong Kong, with an Oxford Minnie Minor, and successfully passed the test:

Hong Kong Driving Licence No.69933 – Issued 6 August 1957.

§

Preparation for Eilish's Arrival in Hong Kong

Further to arrangements made by me earlier in the month, head office confirmed by a letter delivered to me in Sandakan dated 25 September 1957, that London Office had completed travel arrangements for Miss Hannigan to travel to Hong Kong. Booking had been made on BA Flight 934, ETA Hong Kong 31 October 1957. What joy I felt, but I was also acutely aware of the enormous preparations lying ahead of me, and was thankful that Eilish would soon be helping me with final details.

On October 8, 1957, I signed off the *Hew Sang* and was placed ashore on reserve the next day. This was to allow me to prepare to receive Eilish when she arrived from London three weeks later, on October 31.

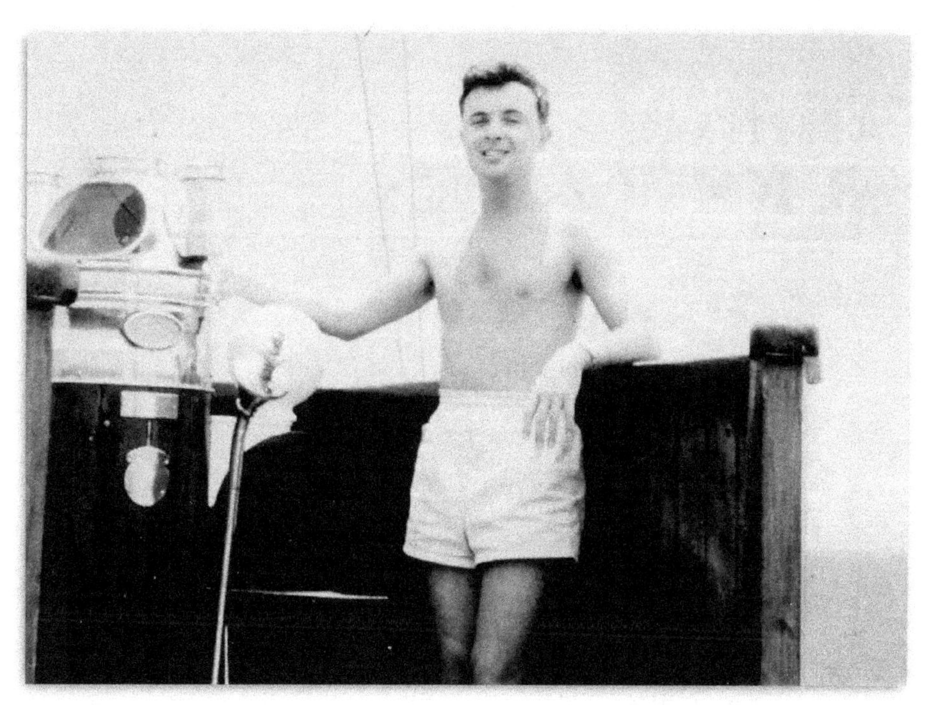

Me getting some sun on the "Monkey Island" of the Hew Sang at sea.

Margaret on trip with husband George Hurford my Second Engineer. c1957

CHAPTER 14:
A NEW LIFE

Marriage and Happiness

1957 – 1958

Finally, the big day came when Eilish and I were to see each other after nearly 18 months' separation. Julie and her husband drove me to meet Eilish on her arrival at Kai Tak Airport on October 31; it was a beautiful sunny day. I was so looking forward to this day, but a little nervous at the same time, as no doubt she must have also felt, on top of her fatigue from the flight. My lovely bride-to-be met us all with a radiant smile, and from that moment, I knew everything was going to be all right. Of course, we all exchanged hugs!

The plan was that Eilish was to stay with Julie at Nairn House until the day of the wedding. In the meantime, I would remain in the European YMCA next to the Peninsula Hotel. It was understood that in the days in between, Eilish and I would be very busy finalizing plans.

With the wedding less than a week away, the next few days were quite hectic. We had to meet our priest at St. Teresa Church on Prince Edward Road. Then on to the Registrar's Office to formalize the civil marriage. Next to Sennett Freres, the jewellers, to select the engagement ring and wedding bands. Julie was going to look after the wedding cake and prepare her place for the reception. I ordered a supply of wine from Maxie Jones, Jardines' procurement superintendent. Before Eilish's arrival, I had already booked a hotel in Macau. Julie was a great help as Eilish adjusted to the new time zone and climate change and got acquainted with the family.

We also took time to briefly look over the flat Jardines had allotted us, on the second floor at 131 Boundary Street. We were quite pleased with what we saw, but to fill the empty rooms would be a daunting task for us. Eilish's own wardrobe included only what she was permitted to check in on the flight, so she and Julie spent some time shopping to fill in the gaps.

Poor Eilish was making the biggest adjustment in her whole life without the support of family, friends or even her parents, and it was only much later that I fully appreciated how bewildering it must have been for her.

Wednesday, November 6 was a beautiful autumn day, with hardly any clouds in the blue sky. My best man Bill Bennett and I dressed in uniform arrived at the church; Eilish, dressed in a beautiful wedding gown custom-made in Ireland, was led by Cecil Renfrew, my brother-in-law, standing in for her father. Julie was the matron of honour, my five-year-old nephew Roger was a page, and three-year-old Robyn was a bridesmaid. It was a beautiful setting.

The exchanging of vows, presided over by the priest, was flawless. I was delighted to see so many familiar faces in the pews witnessing this joyous occasion, and as Eilish and I walked down the aisle, George Wallett wise-cracked 'another good man gone west' with a smile.

No formal announcement or invitations were sent out to my friends, as most of them were away somewhere on their ships. However, those who happened to be in port honoured me by their attendance with their wives. Everybody who came to the church to honour us was automatically welcomed at the reception held at the Renfrew's' residence at Nairn House. Bill and Pat Bennett came, of course, and so

Wedding party at church steps. November 7, 1957

Bridal group exiting the church under shower of confetti.

Bob, Eilish and the wedding cake at the reception.

Maxie Jones and his wife, Mavis Bartlett, Pat Bennett, and Mike Pope at the reception. 1957

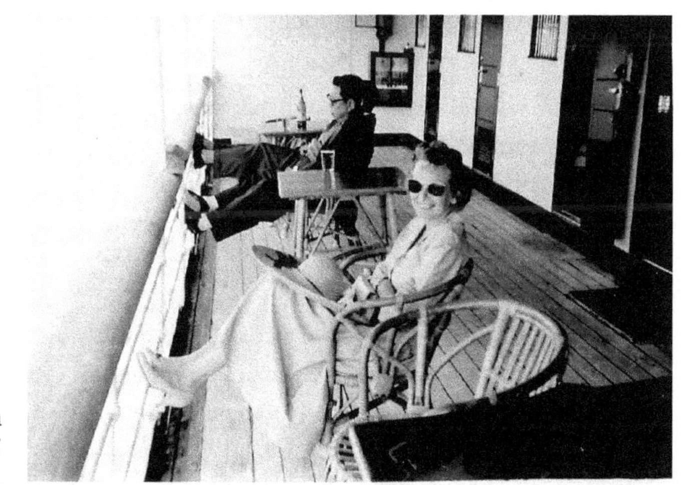

Eilish on ferry boat to Macau on honeymoon 1957

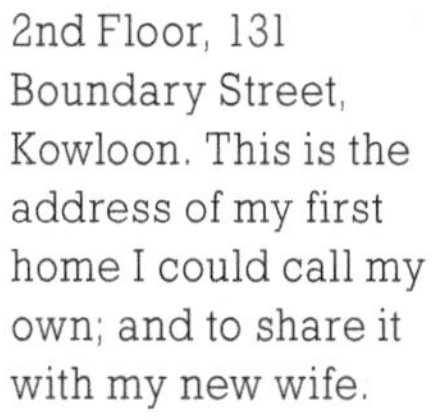

2nd Floor, 131 Boundary Street, Kowloon. This is the address of my first home I could call my own; and to share it with my new wife.
Source: Gwulo.com c1957

did Maxie Jones and his wife, George and Margaret Wallett, Dorothy Tipple and her daughter Lesley, Jan Jensen and his wife Joan, Mavis Bartlett, George and Margaret Hurford, Mike Pope and his wife, and Chas Martin. There were others unknown to me who were friends of my sister and her husband.

At the reception after the wedding, Bill Bennett read all the telegrams, congratulatory cards, telegrams and messages received from well-wishers, and the party warmed up nicely. Eilish very graciously mingled with the guests and was also very attentive to the page and bridesmaid in the house. In keeping with our plan to leave for our honeymoon in Macau that evening, we had to cut short our presence at the reception.

A quick change and we were off by taxi to board the ferry to Macau, departing from the berth in Hong Kong. What a relief it was to at last be alone with each other and let all the tension of the day gradually seep away from our bodies and minds. Gazing upon Eilish as she relaxed on the ferry, looking over the waters, I was awed to realize that she was now my wife and that I was a married man! I was, happy, scared, confused and worried, all jumbled together.

This was our first experience in Macau, a Portuguese colony, smaller in size than the island of Hong Kong. Its claim to fame is its casinos (in Hong Kong, it was against the law to gamble) and for being a popular honeymoon destination. The Hotel Isabel, recommended by Reggie Xavier, was a charming old colonial-style hotel right by the water's edge. Our large suite included a verandah overlooking a muddy Pearl River estuary, where we enjoyed our breakfast every morning. Getting to know each other was heavenly bliss, the only shadow being that inevitably we would both face separation again due to my work. After eight days, we took the ferry back home, tired and refreshed, but completely satisfied.

§

Back home in Hong Kong, we settled down to paying our bills and sending out thank you notes. Getting to know our servant was a first big step for Eilish, and I told Eilish that her title was 'Missy', and mine was 'Master'. She should soon learn how convenient her servant would be. Taking stock of what we needed to furnish our home was an important task; we had already placed some orders, but there was more to come. Eilish urgently needed to augment her wardrobe, but shopping with Julie relieved me of that chore.

This modern building had three flats, all belonging to Jardines. We were located on the second floor and the top floor was occupied by Fred Christie, his wife Maureen and their two sons, Michael and PD (short for Philip David). They were both Australians, and Fred was a deck officer with Jardines.

Our flat was very spacious, and contained a large entrance hallway, large living room, two bedrooms, a dining room, two bathrooms, a kitchen, servants' quarters and a verandah. All the flooring was wood parquet, and each bedroom and the living room had their own coal fireplaces.

As I was not much help due to my inexperience, I counted on my sister to assist Eilish to settle into her new life and quarters. There were also other Jardine officers' wives nearby whom she could contact should she need help. With this knowledge, and knowing what support was available, I felt comfortable leaving her to fend for herself as I prepared to return to work on my next assignment.

SS Tak Sang 太生 *(Vigorous Growing): November 1957 – July 1958*

I was appointed chief engineer on the *Tak Sang* and was very happy that Bob Houghton was my second engineer. Bob and I had served our apprenticeship together at HWD.

The trade route for the Tak Sang was Hong Kong > Singapore > Straits Settlements > Calcutta > Japan and China; the length of the voyage was typically three months, with two opportunities to visit home. The machinery on this ship was new to me, and I relied on Bob Houghton to keep me informed of any issues or problems. On this first trip, the destination was westbound towards Calcutta.

This 3,318-ton ship was built by Short Bros. in 1946, and the 3000hp reciprocating steam engine was built by Northeast Marine. Instead of a steam slide valve, the engine was equipped with a Lentz Poppet valve system. This was new to me. Three oil-fired high pressure boilers supplied steam to the main engine. It was classified as a cargo liner, with 12 first-class accommodations and space for 200 deck passengers.

§

My fears for leaving my new wife, in a strange country and practically on her own, to establish a home, was soon to receive a formidable jolt -- a jolt that would eventually lead to a life-changing decision.

I do not recall from whom I received the news and in what form, but upon arrival in Calcutta, I learned that my wife had been admitted to St. Teresa's Hospital, located across the road from our home; no details were available. Frantically, I telephoned my sister. She confirmed that Eilish had been admitted into hospital with an unknown condition, but according to the doctors, there were no life-threatening concerns.

I immediately telephoned the hospital and spoke to the head nurse, who was a nun. She corroborated what Julie had told me. She also said that the attending doctor, Dr. Wedderburn, intended to perform exploratory surgery to determine the cause of Eilish's extreme retching and nausea, but that I need not be concerned as the nurse would be alongside Eilish throughout the procedure.

My instincts screamed at me to immediately request to be relieved from the *Tak Sang* so that I could take a flight home to be with my wife. I agonized over this dilemma, and eventually concluded that if Eilish's condition was not considered critical, all I would do in the hospital would be sitting and providing personal support to my wife. I eventually compromised: I would not request to be relieved from my ship, I would rely on updates from the hospital should an emergency develop, and Julie would monitor Eilish's condition daily until I returned home in about two or three weeks' time.

By the time I returned to Hong Kong in mid-December, Eilish had been discharged from the hospital and was at home waiting for me. She was weak but in good spirits. Her medical symptoms were constant nausea with frequent vomiting; this was confirmed to be due to her pregnant condition! Of course, the news was thrilling, but I hadn't yet got over the fact that Eilish had just had abdominal surgery as well, and her well-being and comfort were uppermost in my mind. To control the frequent vomiting, she had to visit the hospital just across the street several times a week for b-13 inoculations, and that was her only inconvenience.

It was at this time that I engaged Dr. Christina Chow to attend to Eilish's pregnancy to term. Dr. Chow had been recommended by Reggie Xavier's wife Cynthia, and Eilish felt comfortable with this decision. Reggie had been my second engineer on the Chun Sang, and I knew him well from the days of HWD, where he had also served his apprenticeship.

The night before I returned to the Tak Sang, Eilish gave our servant the night off and insisted she would cook our first meal at home—a mixed grill. The weather was unseasonably cold, so I lit some coal in the living room fireplace, which made the room cozy. To obtain maximum warmth from the fireplace, I had pushed the coffee table very close to the burning fire, and there we ate our dinner by the light from the fireplace and two candles. It was a very romantic evening which refused to be spoiled by a charred coffee table and me pretending to enjoy eating half-cooked kidney for dinner!

Me (l), Wilf Briggs and a deck officer on the Tak Sang. April 1958

Eilish Accompanies Me on a Round Trip to Japan

It was common practice for management to grant permission for wives to accompany their husbands on their ship for one trip per year; I had seen this done for Pat Bennet, Margaret Hurford and Joan Jensen. In due course, I received permission for Eilish to join me, subject to a doctor's approval, on a northern round trip to Japan.

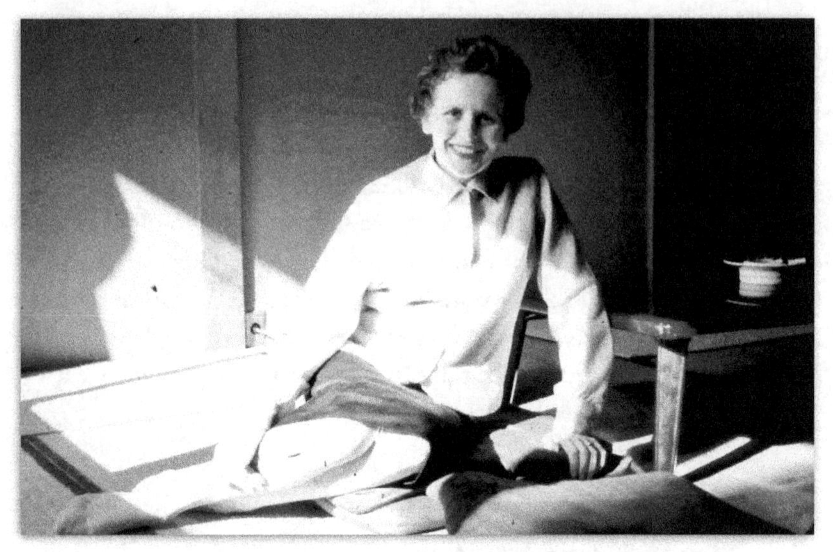

A round trip on the Tak Sang to Japan. We both experienced an overnight stay in a Japanese style hotel in Kyoto. 1958

In April 1958, with her doctor's approval, and on condition that I administer b-13 injections as she needed, Eilish was all set to embark on our second honeymoon on a three-week round trip to Japan. She was feeling well, but her pregnancy was becoming obvious which affected her mobility, so that was all we had to be careful about.

Eilish was rather self-conscious about her condition and the unpredictability of her illness, and because of this, she tended to minimize socializing with the rest of the ship's officers. She was happy to eat all her meals in the confines of my comfortable cabin but was friendly with officers she would meet outside on the deck when enjoying fine weather; of course, she got to know Bob Houghton, my second engineer, well.

With leave from the master (captain) of the ship, we took a train in Kobe to Kyoto and checked into the Imperial Hotel for a couple of nights, which catered mostly to Japanese clientele, so one slept under bedclothes on tatami flooring; this was unexpected but not unpleasant. Eilish, in her sixth month of pregnancy and starting to show, needed help to lower herself to the floor and to get up, but was otherwise unbothered by this experience.

The hotel grounds were magnificent, enhanced by cherry trees in full bloom in different shades of pink and red. We strolled through various areas of the grounds and took many photographs. Nearby these

Return homewards from Japan on the Tak Sang after a brief second honeymoon, together with two budgie birds, each in its own cage. 1958

grounds was the Kinkakuji Temple, beautifully reflected in a body of water.

It was good to have Eilish with me, and it made me dream of a future where new opportunities would present themselves for us to be together. We could get past her illness in Hong Kong and move forward to greater heights in my career with Jardines. In less than two years, I would be eligible for eight months of long leave to spend in Ireland with her parents and siblings, and at that time, decide what else there might be for us in the future. In the meantime, the good life was in Hong Kong.

Before departing Kobe on April 16, we went shopping, and Eilish couldn't resist buying two budgie birds, each in their own cage, to take back with us to Hong Kong.

Our experience in Shanghai was a little different. Soon after our ship was docked alongside our wharf, the political commissar boarded our ship to extend a warm welcome to the passengers and officers. I found him to be a friendly individual, so I invited him to my cabin, and introduced him to Eilish. He also arranged a mini tour of Shanghai in a chauffeured car. We were taken on a short tour of the Bund and took in some of the river traffic scenery before returning to our ship.

Tak Sang departed Shanghai on April 21, 1958 and arrived in Hong Kong on April 27, 1958. With Eilish back in Hong Kong and safely at home, I was to make one more trip to Calcutta before being taken off the *Tak Sang* for my next assignment.

SS Hew Sang for the Third Time: July 1958 – January 1959

The transfer in Hong Kong between Tak Sang and Hew Sang was disappointingly short. I only enjoyed one day and night at home before I was off to North Borneo on the Hew Sang.

The trade route of this ship was ideal for married officers because absence away from home was minimal, and the stay in port was generally better than for ships on other routes. Typically, absence from home was around three weeks for each voyage, and a stay in home port could be up to a week.

I was happy to be back on the Hew Sang as chief engineer, no doubt orchestrated by head office so that I would be close to home when our first child was expected in August. I was familiar with most of the officers, and quickly settled into my former routine.

A Gift from Eilish – A Daughter

I had left home and was at sea on my second trip, heading towards North Borneo, when on August 15th, Chas Martin, our radio officer, delivered a message to me announcing the arrival of my newborn daughter. I was in the middle of a bridge game with three other officers, and we all raised our drink in toasting my wife and baby. I then sent flowers, including congratulations for a job well done and love by wire to my wife.

The introduction to my daughter took place in St. Teresa's Hospital about two weeks later. Eilish was radiant and seemed quite rested and ready to be taken home with baby. With thanks to the doctor and all the nursing staff, the three of us went home to start a new life.

In between voyages, I was going to see a lot of my family, and with the loan of my brother-in-law's car, I was able to show Eilish some of the sights in Hong Kong. I remember one memorable afternoon at the Repulse Bay Hotel, we were having tea when Orson Wells with his four-year-old daughter stopped at our table and he said to her, "Now why can't you be a good little girl like this baby?"; we were quite flattered. Sitting at the next table was his smiling wife, Sylvia Simms.

My sister Julie and family were a great support to Eilish and baby, which was a great relief for me when I was away from home.

Discharge from St. Teresa's Hospital

Eilish, happy with our daughter 1958

CHAPTER 15: SWALLOWING THE ANCHOR

New Horizons And New Dreams

1959

Only about two months had passed, when on my next trip to Japan, I was shocked at the totally unexpected announcement from my wife that she was pregnant again. Even though she displayed no unusual physical distress this time, I was still haunted by memories of her first pregnancy and I feared that at any moment, she would succumb to the same experience again—an irrational conclusion on my part, but nevertheless unshakable.

This was when Eilish and I had a very serious discussion about me seeking a new career that was land-based; I wanted to be close to my wife throughout her pregnancy and to be home for the birth of my next child. This decision would also include a clean break, not only from my seagoing career, but to retire from Hong Kong altogether, a place where I could claim no roots. Australia and the UK were options under consideration, and so was Canada.

The main flaw in my plan was that I failed to first seek professional advice from my boss in head office, John Green, the superintendent engineer, and to seek personal advice from my friend Bill Bennett, the assistant superintendent engineer by this time, who was the best man at my wedding. Maybe it was the severe time constraint facing us, and me being away from home as well, that clouded our decision-making process.

I favoured starting a new life in Canada, a gamble because I knew very little about Canada except from what I had read and heard about in various media outlets. Eilish didn't seem to object but appeared cautious. Furthermore, Ireland was more readily accessible from Canada than Australia, where many Jardines personnel settled.

I applied for immigration to Canada at the office of the Canadian High Commission in Hong Kong as soon as possible. At first, there was a concern about the results shown in Eilish's medical report, resulting in an anxious delay for final approval. Mr. John Armstrong, the immigration officer we dealt with, very kindly expedited our application process for final approval on April 3, 1959.

I was relieved from the Hew Sang on January 12, 1959, and placed on reserve to January 31, when my termination of employment became effective. It was at this time that I had to vacate the company flat on Boundary Street. With immigration finally cleared, Jardines arranged our transportation to Vancouver. All this delay severely impacted our finances.

Our good neighbours Fred Christie, an officer with Jardines, and his wife Maureen put us up in their flat from February 1 to April 5, over two full months. All our furniture had been crated and was in storage during this time. On the day of departure, Maureen and my sister, with her two children, came to see us off on board the *Eastern Saga*, departing for Yokohama.

Jardines paid all my costs from Hong Kong to Vancouver but excluded my wife and daughter because I had not completed my second contract. We travelled on the MS *Eastern Saga* as passengers arriving

Part of my Jardine family at the christening of Hilary Bennett. Elizabeth Stormont holding Hilary, Bill Rowe, Pat Bennett, Norm McQueen, Barbara Taylor, Terry Crichton, children in front: Katie Stormont, Diane Stormont, Cameron Taylor. *Photo by Bill Bennet c1960*

Send off on the *Eastern Saga* in Hong Kong as we start our journey to a new life. Eilish is second from the right.

Arrival into Yokohama on the *Eastern Saga*. Horace, my faithful cabin steward of the past is bidding farewell to my wife and daughter as we get ready to disembark the ship. 1959

in Yokohama on April 15th, where we took the opportunity to visit Tokyo. We next boarded the MS Ventura to continue our voyage across the Pacific Ocean to our destination in Canada, departing Yokohama on April 17, 1959.

After about 11 days at sea, we arrived in Vancouver on April 28, 1959, at the most gorgeous time of the year. The cherry trees were in full bloom and the sky was a clear blue, with puffball clouds scudding overhead. It was magical, and to me, a sign of great encouragement—an omen for new opportunities in a new country. Eilish didn't quite share my optimism, but at least in Canada we were closer to Ireland than we had been in Hong Kong.

What did I give up by making this significant change in our lives? The benefit of eight months with full pay on long leave in Ireland in April 1960 (12 months away); continuing living the lifestyle of expatriates in Hong Kong, including household servants; established seniority and favourable prospects for the future; a generous provident fund to mature at age of retirement at 55, for me that would have been in 1986; a pension plan and medical insurance; and in the meantime, living in subsidized housing in Hong Kong, very likely an elitist residential address somewhere on the Peak..

Getting our sea-legs on the high seas on the Ventura in the Pacific Ocean. 1959

MS Ventura - Yokohama to Vancouver - 1959

AFTERTHOUGHTS

How can I describe all the emotions I experienced over my first 28 years in reaching this point in my life, 21 of which were spent exclusively in Hong Kong? Is it possible to claim a heritage when critical formative years were largely spent in a vacuum? Society in Hong Kong was not very kind toward an orphaned boy of seven years. Subsequent abandonment at 10 was a shocking experience in a time of war, and it was only by the grace of God that I did not become a casualty of the fighting like so many others did. To cap it all, there was no mentorship during the bewildering period of transformation from childhood to adolescence.

The first seven years of my life may be considered normal, since there was some exposure to family life, even though it was fractured. The second part found me stranded alone at the time of my mother's hospitalization with a terminal illness. My formative years in the third part were the saddest years that no child should be made to endure. The fourth part exposed me to the highest physical risks due to war. Emerging from the war in the fifth part, I enjoyed renewed happiness. The sixth part saw me preparing to take control of my own life.

No words can describe the happiness I experienced during the glory years in the seventh part. And finally, in the eighth part, the 'icing on the cake', with my marriage and ending with our arrival into Canada, to start a new life with my own growing family. This is not only a story of survival under great disadvantage, but at the end of it all, also the achievement of significant success.

To make a conscious move to start a new life in a new country, I unconsciously sacrificed promising opportunities for significant advancement and risked losing all my gains. Going into the sunrise, my fervent hope was to be able to give my family a heritage and traditions which I never had—that which they would be able to identify with and grow from.

I knew nothing about Canada. I entertained no anticipation except to check things out first. There were no connections of family, or even former friends. It was a real gamble, requiring a lot of courage and self-confidence.

This story of survival and success was a significant accomplishment under challenging conditions. Success beyond imagination was the ultimate reward for persevering! Was I going to be capable of reproducing this success in my new country of choice? I had no doubts, but I also had no guarantees; in retrospect, a daunting prospect!

Persevering appeared to be my main attribute as I faced many unexpected challenges. It took nearly 20 years in Canada to recover financially. But success did return in my life after I had made the decision to go into business on my own.

In Canada happiness was dealt a severe blow in the loss of my youngest daughter in 1987, followed by the loss of my Irish Rose approximately a year and a half later. I can only thank God for the support I received from my surviving family in Canada, without which life would have been very difficult.

May God give you...
For every storm, a rainbow,
For every tear, a smile,
For every care, a promise,
And a blessing in each trial.
For every problem life sends,
A faithful friend to share,
For every sigh, a sweet song,
And an answer for each prayer.

– Irish Blessing

INDEX

D

E

F

G

H

I

J

K

L

M

N

ACKNOWLEDGEMENTS

I wish to express my gratitude to all those individuals who, in one way or another, contributed significantly to the development of my memoirs, or are a major part of my memories. Communication varied from face to face interviews, via Internet, and importantly also from contributors to David Bellis website www.Gwulo.com.

I am thankful to many of my contemporaries who are still alive. Unfortunately, a few have passed away, but their memory is contained in these memoirs:

Barbara Anslow (née Redwood), a truly legendary person well-known for her diary in Stanley Camp. On my visit to Frinton-On-Sea, Essex, she expressed her interest and encouragement in the development of my story.

Ruth Baker (née Sewell), for her kindness in being interviewed in Oxford and for her encouragement in the development of my story. Also, for her permission to publish the photo of the Bowen Road Group of Children.

David Bellis, of www.gwulo.com, a wonderful source about old Hong Kong, and his response to my numerous enquiries. This is where I followed Tin Hats & Rice by Barbara Anslow.

Olga Blake, nee Robinson, my godmother, who in my adolescent years after the war, encouraged me to pursue a career in marine engineering. To this end, she provided me with sustenance and accommodation for over a period of five years. I benefited enormously from this generous gesture.

Sr. Theresa Chien of the Canossian Institute, who kindly gave me access to History of Our Canossian Missions, Volumes I, II and III, which revealed the fascinating story of Emily Lucy Ann Bowring, youngest daughter of the former governor of Hong Kong (1854-1859). Also, to thank her for her efforts in my search on early family activities. Interviews in Hong Kong.

Philip Cracknell, for so generously and promptly answering numerous questions to help in the development of my story. (Originator www.BattleforHongKong.blogspot.hk)

Mike Cussans, a photographic contributor to www.gwulo.com, for his kind permission to use copies of his photographs of Old Hong Kong. Direct communication by Internet.

Brian Edgar, is an active historical consultant and presently engaged in research on the experience of British civilians during the Japanese occupation of Hong Kong. I value his expertise and appreciate his contribution of the foreword for this book.

Pamela Garneau (née Armstrong, daughter of John Armstrong, the Canadian immigration officer in Hong Kong who processed my application for entry into Canada). Presently resident of Gatineau, Quebec, Canada. Contact made. Lasting memories.

Sr. Huguette Turcotte, Archivist at MIC in Quebec, who kindly provided information on MIC activities in Hong Kong and introduced me to the family of Sr. St. Stanislas (Germaine Gonthier) in Quebec.

Lesley Kilvert (nee Tipple). The Tipple family for pre-war weekend visits in Red Roofs, escorted by Aunty Olga, my godmother, that gave me intense pleasure, with friendship lasting ever since.

Uncle Costia Kriloff and Aunty Lydia, for the frequent pre-war weekend breaks from the boarding school during my early orphan years and those lasting memories.

Caroline Milne (née Bartlett, daughter of 'Bart', with whom I sailed with in Jardines). Mavis attended my wedding in Hong Kong. Contact made. Resident of UK. Internet communication.

Sue Ponsford (née Genders), daughter of Christine Corra of pre-war days in Hong Kong and Stanley Camp. Interesting contact was made. Resident of West Sussex. Lasting memories. Internet communication.

Deborah Povey (daughter of John Povey, with whom I sailed with in Jardines). Resident of UK. Internet communication.

The Rodgers for the pre-war invitation to their fabulous Sunday brunches at 11 Coombe Road, and the affectionate attention from Elizabeth Rodgers and Christine Corra, one of the many guests at the brunch. Lasting memories.

Mother Rosetta, a nun at the Canossian Refugee Centre on Caine Road, who gave me shelter and sustenance until the end of the war. Deceased.

Elizabeth Stormont (wife of John Stormont, director of Jardine, Matheson & Co. in London, and with whom I sailed with in Jardines). Resident of Cheam, Surrey, UK. Lasting friendship ever since.

William Sewell and his family, who took me in at 10 years old at the beginning of the war, and with whose family I found shelter on

Mount Cameron in the Kennedy-Skipton residence during the Battle for Hong Kong and up to our entry into Stanley Internment Camp. Deceased.

Authors and their Works:

With meager family records and limited collection of family photographs, all these readings contributed significantly to the development of this book by stimulating my memory of events from the dim past.

Anslow, Barbara *Tin Hats & Rice:* A diary of life as a Hong Kong prisoner of war, 1941-1945, as first appeared in a serial on www.gwulo.com

Ballantyne, Peter Master Mariner, *Sailing the China Seas: The Indo-China Steam Navigation Company 1881-c1939.* A dissertation submitted towards his MA in Maritime History, University of Greenwich. A very comprehensive history of the origins of Jardine Matheson, and some of their subsidiaries, that has generally been regarded as the heyday of British shipping on the coasts and rivers of China.

Ballard, J. G. *Miracles of Life: Shanghai to Shepperton,* Fourth Estate, 2008. Author of *Empire of the Sun.*

Banham, Tony *Not the Slightest Chance: The Defence of Hong Kong, 1941,* Hong Kong University Press, 2003. This was invaluable in helping me to establish my own timeline of events during the Battle for Hong Kong. Monthly Newsletter: http://www.hongkongwardiary.com/

Birch, Alan & Cole, Martin *Captive Years: The Occupation of Hong Kong, 1941-45,* Heinemann Asia, 1982.

Emerson, Geoff *Hong Kong Internment, 1942-1945,* Hong Kong University Press, 2011. This proved useful in my research for the five months that I was incarcerated in Stanley Camp.

Fortescue, Diana *The Survivors: A Period Piece,* first published by Anima Books, 2015. A family history book, including Stanley Internment Camp.

Greenfield, Nathan E, *The Damned, The Canadians at the Battle of Hong Kong and the POW Experience,* 1941-45, published by Harper Collins 2011

Keswick, Maggie *The Thistle and the Jade: A Celebration of 150 Years of Jardine, Matheson & Co.*, Octopus Books Limited, London, 1982.

Nash, Gary *The Tarasov Saga: from Russia Through China to Australia.* First published by Rosenberg Publishing Pty Ltd., 2002. A moving account of a close-knit family during a tumultuous era.

Parker, Ronald C, *Deadly December, The Battle of Hong Kong,* The Royal Rifles of Canada, The Winnipeg Grenadiers. A dedication to Major Maurice A. Parker

Penlington, Valerie Ann *Winged Dragon: The History of the Royal Hong Kong Auxiliary Air Force,* Odyssey Productions, 1996.

Refo, Sarah (Sally) unpublished account of the invasion of Hong Kong, compiled on August 1, 1942, on board the MS *Gripsholm* during repatriation to the US. This account contains one reference to "an orphan 10-year-old boy from DGS", but nothing more. My thanks to her daughter, Burney Medard, for permission to quote from this account.

Sala, Sr. Ida, F.d.C.C., *History of our Canossian Missions, Hong Kong 1910-2000,* Volume III, chapters 3 to 7.

Selwyn-Clarke, Sir Selwyn, *Footprints: The Memoirs of Sir Selwyn Selwyn-Clarke,* chapters 6 to 9. Sino-American Publishing Co. Hong Kong, 1975.

Sewell, William G. *Strange Harmony,* Edinburgh House Press, London, 1946. In the preface to his book, William Sewell makes clear that his intent was to record a personal account for his family and some of the people they encountered, not a comprehensive history of the Stanley Internment Camp. For this reason, all the characters other than the author's family are syntheses of real people; no character corresponds to a single real person.

Stewart, Evan, *Hong Kong Volunteers in Battle,* Published for RHKR (The Volunteers) Association Ltd., by Blacksmith Books, Hong Kong 2004 and 2005.

Williams, Stephanie *Olga's Story,* Doubleday Canada, 2005. Set against the backdrop of two revolutions and two world wars, and crossing three continents, Olga's Story is the true saga of one woman's remarkable courage and resilience in the face of some of the twentieth century's gravest human tragedies.

Wright-Nooth, George, *Prisoner of the Turnip Heads, The Fall of Hong Kong and Imprisonment by the Japanese.* Cassell Edition 1994

To order more copies of this book, find books by other Canadian authors, or make inquiries about publishing your own book, contact PageMaster at:

ABOUT THE AUTHOR

Having "swallowed" the anchor, a nautical term for abandoning a career on ocean ships, Bob Tatz eventually settled in the province of Alberta in Canada. With credentials of a professional marine engineer it wasn't long before Bob established a reputation as a versatile engineering consultant. He remains a Fellow of the Institute of Marine Engineering, Science, and Technology (UK).

His varied experience in the marine engineering field stood him in good stead and was instrumental for his advancement in his new career. By meeting Canadian and U.S. standards, Bob enhanced his ability to provide a valued service to his clients. With encouragement from loyal industry connections, Bob eventually established his own business, which grew as his client base increased.

Bob is well travelled nationally as well as internationally, both in business and in retirement, and he easily enjoys the opportunity to interact with people from all walks of life.

In writing these memoirs Bob made a great effort to recall some of the most interesting experiences in his early life before immigrating to Canada. With established roots in Canada, Bob is grateful for his successes and wishes to thank all those who made it possible.

For relaxation Bob enjoys art sketching, and looks forward to resuming creating with oil paint again after a long absence.

When we had a "tough" inspection job to confirm pressure integrity in plant or field equipment we could ask Bob Tatz to attend to it. He would always please the operations people, win the day, make more friends and return with a quality detailed report. – **Bruce Fowlie**, *Calgary, Imperial Oil's Civil Engineering Inspection Group*

Bob and I worked together at Chubb Insurance. Bob's knowledge and professionalism are beyond question. However, it is his friendship that I will always cherish. – **Suet Chan**, *Senior Boiler & Machinery Underwriter*

If I ever needed pressure vessels and equipment inspected with efficiency and accuracy Bob Tatz was the first person I would call. I always obtained his complete commitment. Great professional. – **John Povey**, *Imperial Oil*

It is my great pleasure to regard Bob Tatz as a friend as well as a coworker. Not only did Bob bring this wealth of knowledge to his work but was more than willing to share with the next generation. – **Jack O'Brien**, *Chubb Insurance Senior Risk Engineering Specialist*

CPSIA information can be obtained
at www.ICGtesting.com
Printed in the USA
LVHW082336270919
632587LV00014B/1584/P